Number 147
Fall 2015

New Directions for Evaluation

Paul R. Brandon
Editor-in-Chief

S0-BZB-923

Monitoring and Evaluation of Climate Change Adaptation: A Review of the Landscape

Dennis Bours
Colleen McGinn
Patrick Pringle
Editors

MONITORING AND EVALUATION OF CLIMATE CHANGE ADAPTATION: A REVIEW OF THE LANDSCAPE
Dennis Bours, Colleen McGinn, Patrick Pringle (eds.)
New Directions for Evaluation, no. 147
Paul R. Brandon, Editor-in-Chief

Microfilm copies of issues and articles are available in 16mm and 35mm, as well as microfiche in 105mm, through University Microfilms Inc., 300 North Zeeb Road, Ann Arbor, MI 48106-1346.

New Directions for Evaluation is indexed in Academic Search Alumni Edition (EBSCO Publishing), Education Research Complete (EBSCO Publishing), Higher Education Abstracts (Claremont Graduate University), SCOPUS (Elsevier), Social Services Abstracts (ProQuest), Sociological Abstracts (ProQuest), Worldwide Political Science Abstracts (ProQuest).

NEW DIRECTIONS FOR EVALUATION (ISSN 1097-6736, electronic ISSN 1534-875X) is part of The Jossey-Bass Education Series and is published quarterly by Wiley Subscription Services, Inc., A Wiley Company, at Jossey-Bass, One Montgomery Street, Suite 1200, San Francisco, CA 94104-4594.

SUBSCRIPTIONS for individuals cost $89 for U.S./Canada/Mexico/international. For institutions, $358 U.S.; $398 Canada/Mexico; $432 international. Electronic only: $89 for individuals all regions; $358 for institutions all regions. Print and electronic: $98 for individuals in the U.S., Canada, and Mexico; $122 for individuals for the rest of the world; $430 for institutions in the U.S.; $470 for institutions in Canada and Mexico; $504 for institutions for the rest of the world.

All issues are proposed by guest editors. For proposal submission guidelines, go to http://www.eval.org/p/cm/ld/fid=48. Editorial correspondence should be addressed to the Editor-in-Chief, Paul R. Brandon, University of Hawai'i at Mānoa, 1776 University Avenue, Castle Memorial Hall Rm 118, Honolulu, HI 96822-2463.

www.josseybass.com

Cover photograph by ©iStock.com/Smithore

Editorial Policy and Procedures

New Directions for Evaluation, a quarterly sourcebook, is an official publication of the American Evaluation Association. The journal publishes works on all aspects of evaluation, with an emphasis on presenting timely and thoughtful reflections on leading-edge issues of evaluation theory, practice, methods, the profession, and the organizational, cultural, and societal context within which evaluation occurs. Each issue of the journal is devoted to a single topic, with contributions solicited, organized, reviewed, and edited by one or more guest editors.

The editor-in-chief is seeking proposals for journal issues from around the globe about topics new to the journal (although topics discussed in the past can be revisited). A diversity of perspectives and creative bridges between evaluation and other disciplines, as well as chapters reporting original empirical research on evaluation, are encouraged. A wide range of topics and substantive domains is appropriate for publication, including evaluative endeavors other than program evaluation; however, the proposed topic must be of interest to a broad evaluation audience.

Journal issues may take any of several forms. Typically they are presented as a series of related chapters, but they might also be presented as a debate; an account, with critique and commentary, of an exemplary evaluation; a feature-length article followed by brief critical commentaries; or perhaps another form proposed by guest editors.

Submitted proposals must follow the format found via the Association's website at http://www.eval.org/Publications/NDE.asp. Proposals are sent to members of the journal's Editorial Advisory Board and to relevant substantive experts for single-blind peer review. The process may result in acceptance, a recommendation to revise and resubmit, or rejection. The journal does not consider or publish unsolicited single manuscripts.

Before submitting proposals, all parties are asked to contact the editor-in-chief, who is committed to working constructively with potential guest editors to help them develop acceptable proposals. For additional information about the journal, see the "Statement of the Editor-in-Chief" in the Spring 2013 issue (No. 137).

Paul R. Brandon, Editor-in-Chief
University of Hawai'i at Mānoa
College of Education
1776 University Avenue
Castle Memorial Hall, Rm. 118
Honolulu, HI 968222463
e-mail: nde@eval.org

CONTENTS

EDITORS' NOTES

C limate change represents a wicked problem facing policy makers, researchers, and practitioners (Levin, Cashore, Bernstein, & Auld, 2012; Termeer, Dewulf, & Breeman, 2013) insofar as it is deeply complex, intractable, and resistant to solution. Despite increasingly sophisticated climate projections, there is uncertainty over the pace and magnitude of changes and exactly how these changes—themselves mediated by dynamic social, political, economic, and environmental processes—will shape future risks and vulnerabilities. Climate change is already reversing gains made toward sustainable human development in specific localities and may compromise the lives, health, and livelihoods of people across the globe. A United Nations Development Programme (UNDP) (2010) briefing paper, for example, highlighted that climate change represents "a significant threat" (p. 1), undermining advances toward eradicating extreme poverty and hunger. As such, it is an increasingly prominent issue on the international development agenda.

There are two overall thrusts of policy and programming to address climate change: mitigation and adaptation. *Mitigation* seeks to reduce the magnitude or rate of long-term climate change itself, usually through the reduction of greenhouse gas emissions and preserving or restocking forests and other carbon sinks. From a monitoring and evaluation (M&E) perspective, many mitigation activities benefit from an objective, transparent, and universal measure: reduction of greenhouse gas emissions, often measured as carbon dioxide equivalents, a fungible unit enabling comparison between interventions. So although mitigation activities may also need to capture wider social, economic and environmental aspects, there is little or no argument about the importance of measuring reductions in emissions in order to understand the effectiveness of an intervention. *Climate change adaptation* (CCA), by contrast, refers to the adjustment of natural or human systems in response to the actual or expected effects of climate change so that people and systems are more resilient. What precisely that means, however—and how to measure it—has wide interpretations.

As we have written elsewhere (Bours, McGinn, & Pringle, 2014b), M&E of CCA poses a bundle of thorny methodological challenges. Individually, none are unique to CCA, but together they represent a very distinctive conundrum facing practitioners and policy makers. CCA has emerged as

---

Dennis Bours, Colleen McGinn, and Patrick Pringle are also co-authors of a series of highly regarded publications on monitoring and evaluation of climate change adaptation. Their articles can be found at: http://www.ukcip.org.uk/publications/monitoring-evaluation-reports/ or http://www.seachangecop.org/search/node/bours%20mcginn%20pringle

NEW DIRECTIONS FOR EVALUATION, no. 147, Fall 2015   © 2015 Wiley Periodicals, Inc., and the American Evaluation Association.   Published online in Wiley Online Library (wileyonlinelibrary.com) • DOI: 10.1002/ev.20127

a cross-cutting issue across the field of international development. Agencies are now struggling how to define, measure, and demonstrate their achievements—and justify their funding. The initial response was to seek to identify a collection of adaptation indicators as benchmarks, but CCA is, ultimately, a poor methodological fit for universal indicators (Bours, McGinn, & Pringle, 2014a). Climate change may be global, but adaptation is a fundamentally local process that cuts across scales, sectors, and levels of intervention. CCA specialists must explore novel ways to approach program design, monitoring, and evaluation to document achievements and lessons learned, and to contribute to an emerging evidence base.

This journal issue comes at a critical time. It presents findings from many of the most important contemporary CCA program evaluation research initiatives. In recent years, a proliferation of tools and resources within the "grey literature" has been published (see Bours, McGinn, & Pringle, 2014c). However, there has been limited reflection on and across the range of emerging approaches and concepts in academic literature. Consequently, this issue is both timely and much needed, and represents a collection of state-of-the-art analyses that address key aspects and debates in CCA M&E. The articles presented here help advance a developing field of practice that addresses one of the world's most pressing issues by providing a compendium of current perspectives and insights on adaptation M&E, often with a strong focus on practical application. In some cases the authors build upon and expand their earlier work by providing deeper reflection on the application of previously published frameworks and approaches, while in other cases the articles present emerging findings from innovative developments in the field.

We strongly believe this issue will be of interest and relevance to a wider community of M&E specialists and scholars as the specific methodological challenges that characterize CCA M&E are not—at least individually—unique, and can inform strategies of those grappling with these challenges in other fields. There is much ground for cross-fertilization and learning; the effective engagement of stakeholders in evaluation addressed in a number of the chapters provides just one example of the wider applicability of lessons. A second point is perhaps less obvious, but perhaps more important. In an age of globalization, localized issues are driven by global influences and trends (European Environment Agency [EEA], 2015). How then to effect change—and measure it? This would appear to require a careful balance between the collection of coherent sets of data across heterogeneous programs and contexts in order to influence international-level policy making, and the production of more specific learning tailored to local experiences and situations. Given the growing number of truly global factors shaping decisions in many aspects of society, we believe that the need for both globally relevant and locally focused perspectives derived from M&E outputs is likely to increasingly influence M&E in general. Consequently, evaluation practitioners and experts in other programmatic areas stand to benefit from the insights and experiences from cutting-edge CCA evaluation research.

The Challenge of Monitoring and Evaluation of Climate Change Adaptation

Climate change adaptation represents a new focus for development programming, albeit one that often builds upon existent practice in other thematic areas, for example, livelihoods, disaster risk reduction (DRR), natural resource management, and food security. Consequently, there are synergies between CCA and sustainable approaches in these areas, for example, the appreciation of the importance of ecosystems in both livelihoods systems and in enhancing resilience to climate change. On one hand these overlaps present opportunities to converge good practice and, on the other, a risk of repetition and relabeling of projects and programs to fit prevailing funding agendas. Despite these similarities, CCA and "good development" are not necessarily interchangeable as Spearman and McGray (2011) emphasize; "practitioners planning interventions should recognize that not all development is adaptation and not all adaptation leads to development" (p. 11). There is a consensus that for CCA interventions to be effective, they must be tailored specifically to the challenges and dilemmas posed by climate change, while tackling the underlying drivers that generate risk and vulnerability. What precisely that means, however—and how to measure it—has wide interpretations. M&E plays a central role in identifying how best to reduce vulnerability and build resilience to climate change (Bours et al., 2014c), especially when knowledge is shared between and across adaptation projects and programs, and between stakeholders. It can play an important role in enhancing our understanding of the links between CCA and other development themes. For example, M&E can be used to explore the ways in which the assumptions that underpin conventional development are changing to reflect climate change and other megatrends (EEA, 2015) linked to planetary boundaries (Rockström, 2009), such as population growth and biodiversity loss.

A common question raised is: Given the long-established expertise and experience of monitoring and evaluating development activities, why do we specifically need to consider M&E practice in the context of climate change adaptation and resilience? CCA poses challenges of unprecedented scale and scope, which cuts across normal programming sectors, levels of intervention, and time frames. Defining, measuring, and evaluating it is methodologically complex: CCA exhibits a number of characteristics that require specific consideration if monitoring and evaluation is going to be effective (Bours et al., 2014b). Issues addressed by the authors in this volume include the following:

Long Time Horizons

Climate change is a long-term process that stretches far beyond the span of program management cycles. The real impact of CCA interventions may not be apparent for decades. How then to define and measure achievements? (see Karani, Mayhew, and Anderson, Article 5).

Uncertainty About Actual Climate Change Patterns and Their Effects in a Given Locale

Although we are confident that climate change will trigger more severe adverse weather events globally, it is unclear exactly how and when changes will unfold and what their consequences will be in situ (e.g., Faulkner, Ayers, & Huq, Article 6).

Shifting Baseline Data and Changing Contexts

The normal approach to program evaluation includes collecting baseline data against which progress can be tracked. However, climate change impacts are inherently uncertain and occur against a backdrop of complex, dynamic socioeconomic and environmental processes. Comparison of pre- and post-intervention data thus loses validity (e.g., Fisher, Dinshaw, McGray, Rai, and Schaar, Article 1).

Contribution vs. Attribution

M&E approaches often seek to demonstrate that changes can be attributed specifically to a particular endeavor. However, the complexity of climate change adaptation and related interventions require a modified approach to M&E. Implementers instead need to demonstrate how their policy or program contributes to an overall adaptation process that is largely shaped by external drivers (e.g., Leiter, Article 8).

Mismatches Between Divergent Adaptation Values, Perceptions, and Goals

Different stakeholders—and technical experts—are likely to make differing judgments and come to alternative conclusions. This can result in conflict or simply sidelining one or more critical perspectives (e.g., Krause, Schwab, & Birkmann, Article 2). Although not unique to CCA, this challenge is especially pertinent given the inherent uncertainties, long time horizons (including intergenerational responsibilities), and complexities of adaptation.

Measuring Complex, Multi-objective, and Interdisciplinary Strategies

Evaluating CCA can be difficult because it crosses sectors and disciplines and reflects multiple influences across the social and natural sciences. In this context, it can be difficult to build a coherent evidence base or establish linear causal relationships (e.g., McKinnon & Hole, Article 3).

Key Themes

Perception, Values and Context

The emergence of tools and frameworks to support effective monitoring and evaluation can, at face value, appear to mask the underlying complexities of adaptation and underplay the importance of the values and perceptions that underpin how "successful adaptation" is defined, and by whom. A number of chapters in this issue explicitly address these often contested definitions of success, exploring how these differences might transform approaches to assessing the performance of an intervention. By doing so, they challenge M&E practitioners to consider these issues more comprehensively in the future.

Krause et al. take an actor-oriented, context-specific perspective, which is applied to the design of a CCA M&E framework in Vietnam. They argue that individual goals and values are important to consider as adaptation actions are part of complex systems and change their meaning and value in relation to other strategies and actors. Instead of attempting to provide an absolute ranking of the most promising adaptation strategies, they provide a framework for understanding the differential meaning of good adaptation across actors, regions, and time scales. Through this approach, the authors provide insights into differing motivations to act, differing values, and the process of negotiating of diverse priorities among different actor groups. The authors call for more deliberate and flexible designs of M&E program components that avoid absolute judgments and reveal distinct and contested meanings of what constitutes good adaptation.

Faulkner et al. also stress the need for more effective consideration of perspectives and values within M&E methodologies. Their article on community-based adaptation (CBA) argues that for M&E of CBA to be effective, it is imperative to consider not simply what works, but also for whom. They stress that what constitutes success is too often determined by top-down processes and interests, and make the case for more nuanced M&E approaches that reflect community perspectives.

Values and perceptions are embedded within context, thus can shape, and be shaped by, differing interpretations of whether, and under what circumstances, adaptation interventions work. Impacts and vulnerabilities are ultimately specific to local contexts (Chong, Gero, & Treichel, Article 7). This places an obligation on M&E practitioners to appreciate, understand, and examine contextual specificities, even while drawing out key lessons for broader praxis. The Fisher et al. (Article 1) review of existing CCA M&E methodologies stresses the significance of context in determining whether a given method or approach is appropriate. This point is ably illustrated by Adler, Wilson, Abbott, and Blackshaw (Article 4) who describe tools designed to assess the impact of U.K. climate finance in Ethiopia that were tailored to the local context. Fisher et al. make a strong case for a holistic approach using several methods iteratively and pragmatically, which would appear to provide greater flexibility and responsiveness to context.

The breadth of topics examined in the chapters present a further question regarding context: To what extent does, or should, thematic context influence evaluation methodologies? Articles 3 and 7 provide interesting perspectives in this regard, presenting methodologies tailored to specific contexts. McKinnon and Hole (3) consider the emerging field of ecosystems-based adaptation, and Chong et al. (7) focus on methods for a specific group, in this case children and youth. Krause et al. (Article 2), meanwhile, make a persuasive case for cultural competence within M&E, while Faulkner et al. (Article 6) emphasize how top-down, donor-driven M&E agendas push aside critical questions regarding who benefits from adaptation. The examples in this issue provide a valuable reminder to M&E practitioners and funders that embracing rather than avoiding complexity can deepen our understanding of adaptation progress and performance.

The Practical Application of Tools and Methods

As donors and practitioners seek to navigate their way through the range of challenges that climate adaptation M&E presents, so a growing array of tools, methods, and frameworks proposed has flourished (Bours et al., 2014c). Many of the articles in this issue utilize and comment on these approaches in depth. Here, it is worth briefly reflecting on what these articles tell us about practical experiences of applying them, and the common lessons that can be distilled.

Fisher et al. (Article 1) review existing methods used in M&E, and consider their applicability to inherently complex interventions such as climate change adaptation. Their review presents a timely reminder to those designing M&E systems that much can be learned from other fields and that it is not necessary or desirable to reinvent the wheel. They suggest that we should be considering the blend of methods employed in a given context; CCA presents methodological challenges, but given the heterogeneity within the field, there is no one-size-fits-all solution to addressing them. The editors support the view that learning lies at the heart of M&E, and that methods should be selected judiciously. A critical assessment of *what* we want to learn and *why* is required before we can consider *how*? Climate change is not the only wicked or complex problem we face, and its specialists should be alert to opportunities to share methods wherever possible. The role of a multidisciplinary viewpoint on adaptation *within* evaluation frameworks is also highlighted by Krause et al. Their approach also appears consistent with the pragmatic suggestion to mix methods and triangulate evidence presented by Fisher et al. Similarly, McKinnon and Hole recognize the benefits of utilizing both theory of change (ToC) and evidence synthesis in complementary ways.

Some of the chapters in this issue also explore the translation of M&E theory into practice, and provide valuable insight into constraints in implementation, often in challenging contexts where data is limited. Karani et al. present a case study applying the Tracking Adaptation and Measuring Development (TAMD) framework in Kenya. They found that concepts and language needed to be simplified and adapted to context; meanwhile Adler et al. provide useful perspectives on managing trade-offs between practicality (e.g., reducing the time burden on local facilitators) and rigor (in this case the number of capacity components that would be considered). It is clear to the editors that pragmatism is required when applying innovative M&E frameworks and tools, but that this must not be used as an excuse for failing to address the challenges associated with adaptation M&E.

There is no single metric for adaptation (Brooks et al., 2011), so the selection and application of indicators in adaptation M&E almost inevitably comes to the fore. Looking across the nine articles, indicators were developed in different ways; however, there is a common appreciation of the need to complement quantitative indicators with qualitative information, either through the use of more nuanced indicators or simply by recognizing that quantitative indicators must be analyzed and interpreted alongside qualitative information. An example of the former can be found in Chong et al., who base their indicator set on the project's ToC and focus on qualitative change indicators, an approach that captures children's perspectives on resilience in a way that conventional quantitative proxy indicators would not. Leiter, meanwhile, considers the challenges in linking CCA M&E across scales. He identifies the need for national-level indicators to make more effective use of subnational information, not only through the use of standardized indicators, but through more effective sharing of evidence.

A rigid application of theoretical or top-down frameworks will not work equally well in diverse contexts, and will ultimately constrain the effectiveness of M&E methods. In turn, it will limit what we learn about the adaptation itself, and ultimately compromise our ability to cope with climate change. Together, these articles illustrate that a focus on learning, a pragmatic and context-relevant application of mixed methods, and a more nuanced approach to the selection and interpretation of indicators can contribute to more practical and successful M&E praxis. The challenge is for funding organizations to balance an understandable desire for methodological consistency across their portfolio of adaptation investments with the need to allow agile methods to evolve that better cope with the complex and context-specific realities of the field.

Utilizing Evaluation Research to Inform Policy and Praxis

What constitutes "good" adaptation, and moreover what policies should be set in place to advance this aim? Evaluation research constitutes a widespread body of applied research, but is often poorly utilized, and lessons are not fully applied. This is especially problematic in the case of CCA, which is characterized by a diverse proliferation of activity and a weak evidence base. At the global level, it is critical to have coherent data to inform policy. In this volume, Roehrer and Kouassi (Article 9) describe the process by which one of world's biggest climate finance funds developed a more coherent M&E framework, based on a selection of five core indicators and six optional indicators, and an ongoing, iterative, and participatory process for evaluating progress against indicator benchmarks. Adler et al., meanwhile, unpack the methodological challenges regards assessing the impact of climate funding from the United Kingdom, and how it has influenced Ethiopia's ambitious policy framework. They specifically examine the application of tools designed to measure institutional structures and capacities—findings that have broad applications beyond CCA per se, particularly surrounding civil service reform, and how to mobilize local governments around new topics.

Some of the most valuable innovations within CCA M&E have emerged from the micro level; several articles in this issue that reflect inventive and incisive approaches drawn from community-based programs testify to that. However, too often global analysts and specialists who work at the local level have not fully engaged with each other, and lessons have been lost. Voices across this issue highlight that global versus local M&E can be something of a false dichotomy, however, and there is intense interest at all levels in how to effectively bridge M&E systems across scales and disciplines in a way that is both reliable and valid. Karani et al., for example, present a case study from Kenya that explores two-track M&E systems that reflect and tie together national and local perspectives to guide policy and planning more effectively. They also demonstrate how CCA M&E approaches can facilitate mainstreaming M&E into development planning processes.

M&E can and should inform policy and praxis, and be better harnessed to do so. This volume includes many examples of initiatives aimed at doing just that. There is a growing consensus within CCA that what distinguishes it from business-as-usual development programs is underlying analysis of climate change impacts and the social, economic, and environmental drivers of population vulnerability and resilience (Bours et al., 2014a). An argument can also be made that climate change is one of several global phenomena that is challenging and, potentially, undermining dominant international development paradigms. In this sense, evaluation research can serve to unpack not only what constitutes successful adaptation, but what is meant by successful development. Several chapters demonstrate the utility of participatory approaches, challenging assumptions about who defines and measures adaptation—and ultimately has a voice in policy making. In this volume, McKinnon and Hole present a particularly important ground for innovation. The emerging framework of ecosystem-based adaptation emphasizes cobenefits between biodiversity conservation and CCA, and in turn how to advance evidence-based policy making. Faulkner et al. highlight nuanced M&E approaches that enhance accountability and learning across multiple levels of stakeholders. They also directly confront the challenge of how to translate local insights into generalized policy guidance. Like Karani et al., they embrace a multitrack approach that is tailored to the needs and interests of diverse stakeholders. Faulkner et al. also emphasize the need to scale up successful local models and engage policy makers at all levels, while also acknowledging knowledge boundaries and power symmetries in doing so. Leiter squarely confronts the challenge of tying together the diversity of CCA, and explores avenues to build a coherent evidence base to inform policy decisions.

Conclusion

This journal issue presents a diverse set of approaches, innovations and reflections in response to the wicked problem of M&E for CCA, authored by many leading researchers and experts in this field. The key methodological challenges inherent in M&E for CCA that we outline above have parallels in, and relevance to, the evaluation of other complex social and environmental issues. Consequently, as well as advancing CCA, we believe this issue has important messages for M&E practitioners in general. Although there is a strong emphasis on the Global South, the examples and insights in formulating coherent responses to methodological challenges are highly relevant to M&E scholars and practitioners globally, including those in North America and Europe. We have outlined three key themes that emerged across the volume as a whole: the role of perception, values, and context; practical application of tools and methods; and the utilization of evaluation research to inform policy and praxis. We hope that these themes, and others explored in individual articles, can catalyze much-needed exchange between evaluators, researchers, and policy makers, leading to more effective adaptation and improved M&E practice.

References

Adler, R., Wilson, K., Abbot, P., & Blackshaw, U. (2015). Approach to monitoring and evaluation of institutional capacity for adaptation to climate change: The case of the United Kingdom's investment to Ethiopia's climate-resilient green economy. In D. Bours, C. McGinn, & P. Pringle (Eds.), *Monitoring and Evaluation of Climate Change Adaptation: A Review of the Landscape.* New Directions for Evaluation, 147, pp. 61–74.

Bours, D., McGinn, C., & Pringle, P. (2014a). *Evaluation review 2: International and donor agency portfolio evaluations: Trends in monitoring and evaluation of climate change adaptation programmes.* Phnom Penh, Cambodia: SEA Change Community of Practice, and Oxford, United Kingdom: UK Climate Impacts Programme (UKCIP). Retrieved from http://www.ukcip.org.uk/wordpress/wp-content /PDFs/UKCIP-SeaChange-MandE-ER2-donor-agencies.pdf

Bours, D., McGinn, C., & Pringle, P. (2014b). *Guidance note 1: Twelve reasons why climate change adaptation M&E is challenging.* Phnom Penh, Cambodia: SEA Change Community of Practice, and Oxford, United Kingdom: UK Climate Impacts Programme (UKCIP). Retrieved from http://www.ukcip.org.uk/wordpress/wp-content /PDFs/MandE-Guidance-Note1.pdf

Bours, D., McGinn, C., & Pringle, P. (2014c). *Monitoring and evaluation for climate change adaptation: A synthesis of tools, frameworks and approaches* (2nd ed.) Phnom Penh, Cambodia: SEA Change Community of Practice, and Oxford, United Kingdom: UK Climate Impacts Programme (UKCIP). Retrieved from http://www .ukcip.org.uk/wordpress/wp-content/PDFs/SEA-Change-UKCIP-MandE-review-2nd- edition.pdf

Brooks, N., Anderson, S., Ayers, J, Burton, I., & Tellam, I. (2011). *Tracking adaptation and measuring development (TAMD).* (Climate Change Working Paper No. 1). London, United Kingdom: International Institute for Environmental Development (IIED). Retrieved from http://pubs.iied.org/pdfs/10031IIED.pdf

Chong, J., Gero, A., & Treichel, P. (2015). What indicates improved resilience to climate change? A learning and evaluative process developed from a child-centered, community-based project in the Philippines. In D. Bours, C. McGinn, & P. Pringle (Eds.), *Monitoring and Evaluation of Climate Change Adaptation: A Review of the Landscape.* New Directions for Evaluation, 147, pp. 105–116.

European Environment Agency (EEA). (2015). *European environment—State and outlook 2015: Assessment of global megatrends.* Copenhagen, Denmark: Author. Retrieved from http://www.eea.europa.eu/soer-2015/global/action-download-pdf

Faulkner, L., Ayers, J., & Huq, S. (2015). Meaningful measurement for community-based adaptation. In D. Bours, C. McGinn, & P. Pringle (Eds.), *Monitoring and Evaluation of Climate Change Adaptation: A Review of the Landscape.* New Directions for Evaluation, 147, pp. 89–104.

Fisher, S., Dinshaw, A., McGray, H., Rai, N., & Schaar, J. (2015). Evaluating climate change adaptation: Learning from methods in international development. In D. Bours, C. McGinn, & P. Pringle (Eds.), *Monitoring and Evaluation of Climate Change Adaptation: A Review of the Landscape.* New Directions for Evaluation, 147, pp. 13–35.

Karani, I., Mayhew, J., & Anderson, S. (2015). Tracking adaptation and measuring development in Isiolo County, Kenya. In D. Bours, C. McGinn, & P. Pringle (Eds.), *Monitoring and Evaluation of Climate Change Adaptation: A Review of the Landscape.* New Directions for Evaluation, 147, pp. 75–87.

Krause, D., Schwab, M., & Birkmann, J. (2015). An actor-oriented and context-specific framework for evaluating climate change adaptation. In D. Bours, C. McGinn, & P. Pringle (Eds.), *Monitoring and Evaluation of Climate Change Adaptation: A Review of the Landscape.* New Directions for Evaluation, 147, pp. 37–48.

Leiter, T. (2015). Linking monitoring and evaluation of adaptation to climate change across scales: Avenues and practical approaches. In D. Bours, C. McGinn, & P. Pringle (Eds.), *Monitoring and Evaluation of Climate Change Adaptation: A Review of the Landscape.* New Directions for Evaluation, 147, pp. 117–127.

Levin, K., Cashore, B., Bernstein, S., & Auld, G. (2012). Overcoming the tragedy of super wicked problems: constraining our future selves to ameliorate global climate change. *Policy Sciences, 45*(2), 123-152. Retrieved from: http://link.springer.com/ article/10.1007/s11077-012-9151-0

McKinnon, M. C., & Hole, D. G. (2015). Exploring program theory to enhance monitoring and evaluation in ecosystem-based adaptation projects. In D. Bours, C. McGinn, & P. Pringle (Eds.), *Monitoring and Evaluation of Climate Change Adaptation: A Review of the Landscape*. New Directions for Evaluation, 147, pp. 49–60.

Rockstrom, J. (2009). Planetary boundaries: Exploring the safe operating space for humanity. *Ecology and Society*, *14*(2), 32. Retrieved from http://www.ecologyandsociety.org/vol14/iss2/art32/

Roehrer, C., & Kouadio, K. E. (2015). Monitoring, reporting, and evidence-based learning in the Climate Investment Fund's Pilot Program for Climate Resilience. In D. Bours, C. McGinn, & P. Pringle (Eds.), *Monitoring and Evaluation of Climate Change Adaptation: A Review of the Landscape*. New Directions for Evaluation, 147, pp. 129–145.

Spearman, M., & McGray, H. (2011). *Making adaptation count: Concepts and options for monitoring and evaluation of climate change adaptation*. Bonn, Germany: Deutsche Gesellschaft für Internationale Zusammenarbeit (GIZ), Bundesministerium für wirtschaftliche Zusammenarbeit und Entwicklung (BMZ), and Washington, DC: World Resources Institute (WRI). Retrieved from http://www.wri.org/publication/making-adaptation-count

Termeer, C., Dewulf, A., & Breeman, G. (2013). Governance of wicked climate adaptation problems. In J. Knieling & W. Leal Filho (Eds.), *Climate change governance*. Heidelberg, Germany: Springer Verlag.

United Nations Development Programme (UNDP). (2010). *Millennium development goals and climate change adaptation: The contribution of UNDP-GEF adaptation initiatives towards MDG1*. New York, NY: Author. Retrieved from http://www.undp.org/content/dam/aplaws/publication/en/publications/environment-energy/www-ee-library/climate-change/the-contribution-of-undp-gef-adaptation-initiatives-towards-mdg1/17463_UNDP_GEF_MDGi1.pdf

Dennis Bours
Colleen McGinn
Patrick Pringle
Editors

DENNIS BOURS *works as extended-term consultant for the Global Environment Facility's Independent Evaluation Office, with a focus on the monitoring and evaluation of adaptation and resilience. Dennis is liaison representative to the ISO Sub-Committee 7 on GHG management and related standards, steering committee member of the Climate Knowledge Brokers group, and adaptation M&E advisor to the Kresge/ND-GAIN Urban Adaptation Assessment project. Dennis holds an MSc in technology transfer and sustainable development, and a second MSc with a focus on climate risk management and risk indicators in global supply chains. Contact: dbours@thegef.org, dpbours@yahoo.com*

COLLEEN MCGINN *is an independent research consultant based in Phnom Penh, Cambodia. She is an expert in population coping, adaptation, and resilience and has a long track record in applied research, M&E, and program management across Asia, Africa, and the Balkans. Colleen holds a PhD from the Columbia University School of Social Work, an MA in Development Studies from Tulane University, and a BA in Political Science from Ohio University. Contact: cm2530@caa.columbia.edu, colleenmcginn@hotmail.com*

PATRICK PRINGLE *is deputy director at the UK Climate Impacts Programme (UKCIP) based at the University of Oxford. He has led, and contributed to, a range of climate adaptation projects in the UK, Europe, Africa, and Asia. Patrick leads UKCIP's work and publications on Climate Adaptation M&E and his previous experience includes working on major program evaluations for World Bank, IFC, U.K. Government, and regional and local authorities. Patrick is currently leading a study of national-level monitoring and evaluation practice in Europe for the European Environment Agency (EEA) and recently developed Adaptation M&E guidance for the coffee sector. Contact: patrick.pringle@ukcip.org.uk, paddypringle@yahoo.co.uk*

NEW DIRECTIONS FOR EVALUATION • DOI: 10.1002/ev

Fisher, S., Dinshaw, A., McGray, H., Rai, N., & Schaar, J. (2015). Evaluating climate change adaptation: Learning from methods in international development. In D. Bours, C. McGinn, & P. Pringle (Eds.), *Monitoring and evaluation of climate change adaptation: A review of the landscape. New Directions for Evaluation, 147,* 13–35.

1

Evaluating Climate Change Adaptation: Learning From Methods in International Development

Susannah Fisher, Ayesha Dinshaw, Heather McGray, Neha Rai, Johan Schaar

Abstract

This article reviews evaluation methods used in the field of international development to draw lessons for the specific challenges of evaluating climate change adaptation. The three specific challenges identified in climate change and resilience monitoring and evaluation are: assessing attribution, creating baselines, and monitoring over long time horizons. This article highlights a range of methods that can be used in climate change adaptation and concludes that, although the methods are available, it is how they are applied that can help address these particular challenges. Methods used within an overarching conceptual framework that emphasizes mixed methods, participatory methodologies, and an iterative, learning focus can start to address the inherent challenges in evaluating responses to an uncertain future climate. This type of approach and application of a set of methods can also be useful in other areas of evaluation, where the outcomes are very long term and socioeconomic trends are extremely uncertain. © 2015 Wiley Periodicals, Inc., and the American Evaluation Association.

T he effects of climate change often challenge progress toward achieving development objectives by altering the underlying natural and social systems in which development takes place. Increasing financial resources are being invested in climate change adaptation (Buchner et al., 2014), but more evidences are needed to shape future investments and to understand how these projects will perform over uncertain and distant climatic futures. One way to address these issues is through robust monitoring and evaluation that generates evidence and learning that is fed back into adaptation practice (Adaptation Committee, 2014). However, results frameworks for adaptation are often project-based, output-oriented, and tend to emphasize spending over results (Independent Evaluation Group of the World Bank [IEG], 2012). In large part, this is due to the challenges of building and assessing resilience to climate change occurring over an uncertain future, both in terms of the changing climate and also in terms of social and economic trends over time. This can lead to a focus on more tangible and therefore measurable outputs for the purposes of reporting and accountability to donors, but they risk being less relevant to assessing actual changes in resilience to climate change over time.

This issue draws on a review of methods and approaches in international development across the sectors of health, natural resource management, agriculture, and work in fragile states and conflict areas to identify methods that address the challenges of monitoring and evaluating climate change adaptation. The review is summarized in Table 1.1 (see Dinshaw, Fisher, McGray, Rai, & Schaar 2014 for a full review). These methods are explored in the context of different types of adaptation interventions—simple, complicated, and complex (Patton, 2011; Zimmerman & Glouberman, 2004; Zimmerman, Lindberg, & Plsek, 1998)—and three challenges for climate change adaptation and resilience (CCAR) monitoring and evaluation (M&E)—assessing attribution, creating baselines, and monitoring over long time horizons. This report was a synthesis of methods. This issue takes this evidence base and the report as the basis for further analysis, showing how adaptation evaluators can move beyond utilizing specific methodologies for M&E and develops an overarching approach to CCAR M&E.

The Challenges of Monitoring and Evaluating Climate Change Adaptation

To determine the most appropriate methods for monitoring and evaluating adaptation, it is useful to distinguish between interventions that have simple, complicated, or complex designs, depending on the degree to which there is agreement and the certainty that specific inputs will lead to specific outcomes. To summarize Patton (2011), simple interventions are those where there is a straightforward logic between inputs, outputs, and outcomes. Complicated interventions may entail multiple components or stakeholders over long time frames. Complex interventions involve fundamental uncertainties, and often disagreement, about the relationship between inputs and outcomes. A causal chain may only become apparent after

Table 1.1. Summary of Evaluation Methods

	Methodology	Definition	Especially applicable to: Simple	Complicated	Complex	Addresses challenge of: Attribution	Baselines/Targets	Long time horizons	Selected example
Social science methods	Surveys	An investigation about the characteristics of a given population by means of collecting data from a sample of that population and estimating their characteristics through the systematic use of statistical methodology.	×	×		×	×		Heltberg, Hossain, and Reva (2012) conducted an extensive survey after the 2008 global food and financial crisis to provide a narrative from the perspective of those who were affected, and to provide feedback on a range of relief schemes that were implemented after the crisis.
	Focus-group interviews	An interviewing technique whereby respondents are interviewed in a group setting, and the interaction between participants is part of the analysis.	×	×		×			International Fund for Agricultural Development evaluated rural development projects in Gambia, Ghana and Morocco (Leeuw & Vaessen, 2009). The methods included desk reviews, quantitative surveys, and focus group discussions with project beneficiaries, control groups, and key informants.

(Continued)

Table 1.1. Continued

		Especially applicable to:			Addresses challenge of:			Selected example
Methodology	Definition	Simple	Complicated	Complex	Attribution	Baselines/Targets	Long time horizons	
Social science methods								
Semistructured interviews with key informants	A one-on-one meeting with a set of questions to be discussed.		×	×	×	×		Used to evaluate the Department for International Development (DFID)–supported project on Community Driven Approaches to Address the Feminization of HIV/AIDS in India (India HIV/Aids Alliance, 2007). The MSC technique helped promote accountability to beneficiaries by keeping program managers in touch with ground realities as well as continuous review and re-alignment of the program assumptions through learning.
Most significant change analysis	Revolves around asking participants or beneficiaries to tell the stories of the most significant change they have experienced through the program (Davies & Dart, 2005).			×	×			

(Continued)

Table 1.1. Continued

		Especially applicable to:				Addresses challenge of:			Selected example
Methodology	Definition	Simple	Complicated	Complex	Attribution	Baselines/Targets	Long time horizons	
Social science methods									
Outcome mapping	Identifying changes in the behaviors of the individuals, groups, and organizations with which program works, rather than changes in the physical variables that may correlate to the development program objectives.		×	×	×			Outcome mapping on an education project in Zimbabwe used boundary partners, that is, the stakeholders the project wished to influence (in this case schools and teachers), in an evaluative process with the project team. The outcome mapping process relies on four methods: monitoring via self-assessment (by program stakeholders); encouraging feedback, reflection, and learning; promoting internal and external dialogue; and following up on unintended effects (Hyse & Ongevalle, 2008).
Limiting factor analysis	Technique to develop a common understanding of the key factors that must be assessed, and if necessary (and possible) managed, for project or program to be viable over the long term.		×	×			×	Gullison and Hardner (2009) have identified a list of limiting factors relevant to a broad range of project types and ecological systems.

(Continued)

Table 1.1. Continued

Methodology	Definition	Especially applicable to:			Addresses challenge of:			Selected example
		Simple	Complicated	Complex	Attribution	Baselines/Targets	Long time horizons	
Scenario building	Generates a set of possible alternative futures ranging from participatory scenarios to modeling data.		×	×		×	×	Feed the Future program is the U.S. government's global hunger and food security initiative that seeks to reduce global poverty and hunger with sustainable development impacts. A tool has also been developed to facilitate target setting by using a series of national data to run scenarios and to set targets (U.S. Agency for International Development, 2012)

Social science methods

(Continued)

Table 1.1. Continued

Methodology	Definition	Especially applicable to: Simple	Especially applicable to: Complicated	Especially applicable to: Complex	Addresses challenge of: Attribution	Addresses challenge of: Baselines/Targets	Addresses challenge of: Long time horizons	Selected example
Experimental design with randomized controls	Compares the treatment group (e.g., program participants) against a randomized control group (e.g., nonparticipants). Statistical techniques used to analyze results.	×			×			Randomized control trials (RCTs) are often used in pharmaceutical trials to understand the effect of a drug compared to a control group.
Quasiexperimental design	Similar to experimental design with different rationales used to assign control groups. But this is undertaken in a nonrandomized way. Can be, for example, through a pipeline approach using groups at different stages of implementation. Analyzed with the use of techniques such as propensity score matching (PSM) and difference in difference.	×	×		×	×		The United Nations Capital Development Fund (UNCDF) in Nigeria, Malawi, Kenya, and Haiti evaluations used new clients, defined as those who had not yet received their first loan or those who had received their first loan but had not yet completed a full loan cycle, as the control group, and older clients, defined by those who had been in the program for at least 20 months, were the treatment group (White, Sinha, & Flanagan, n.d.).

Experiment-related evaluation design

(*Continued*)

Table 1.1. Continued

Methodology	Definition	Especially applicable to:			Addresses challenge of:			Selected example
		Simple	Complicated	Complex	Attribution	Baselines/Targets	Long time horizons	
Modeling	The construction of physical, conceptual, or mathematical simulations of the real world.		×	×			×	As part of the Lower Red River Meadow Restoration Project in Idaho, the project team took an approach that included a pre- and postevaluation (in 1994, and 2000 and 2003, respectively) of the river restoration area with the use of 17 performance indicators, comprising a suite of physical and biological components that interact within the river and wet meadow ecosystems (Klein, Clayton, Alldredge, & Goodwin, 2007).
Econometrics and statistical techniques								

(*Continued*)

Table 1.1. Continued

Methodology	Definition	Especially applicable to:			Addresses challenge of:			Selected example
		Simple	Complicated	Complex	Attribution	Baselines/Targets	Long time horizons	
Econometrics and statistical techniques								
Stochastic baselines	Captures different states that are not captured by different sets of variables but by probability distributions. This model can be useful as it can consider several alternative futures or scenarios.		×	×		×		The Food and Agriculture Policy Research Institute (FAPRI), focusing on the agricultural sector in the United States (Blanco-Fonseca, 2010) used a stochastic model to estimate a more comprehensive baseline that takes into consideration 500 different scenarios that vary in underlying assumptions about factors such as climate, supply and demand, and so on.
Rolling and reconstructing baselines	Collecting baselines during different stages of the program instead of at one time. Reconstructing baselines using secondary administrative data such as national surveys and some practical techniques such as recall and mapping techniques to reconstruct baselines.		×	×		×		The evaluation of Bangladesh Integrated Nutrition Project lacked baseline data to monitor implementation progress. Three separate secondary measurements were used to reconstruct the baseline and create a new comparison group with the use of propensity score matching (World Bank, 2010)

(Continued)

Table 1.1. Continued

Methodology	Definition	Especially applicable to: Simple	Complicated	Complex	Addresses challenge of: Attribution	Baselines/Targets	Long time horizons	Selected example
Econometrics and statistical techniques								
Normalization	Helps standardize data by different trends (unusual or usual) by adjusting the data against these trends and means.		X			X		The evaluation of the Nutrition Care Process (NCP) program used indexing and composite scoring to normalize their evaluation metrics (such as anthropometric data, physical findings data, biomedical data, etc.) that are affected by variations in several health-related aspects. The standardization technique allows arriving at health and disease outcomes by providing comparable data (American Dietetic Association, 2008).
Propensity score matching	Statistical technique used by programs to construct a matched comparison group that has the same propensity to receive the intervention benefits as the treatment group.	X	X		X	X		The Emergency Social Investment Fund of Nicaragua used in 1998 the World Bank's Living Standard Measurement Study (LSMS) data to estimate baselines for project and comparison groups in water and sanitation, health, and education projects. PSM was used to enhance comparisons between the two groups (Pradhan & Rawlings, 2002; World Bank, 2010).

(Continued)

Table 1.1. Continued

Methodology	Definition	Especially applicable to: Simple	Complicated	Complex	Addresses challenge of: Attribution	Baselines/Targets	Long time horizons	Selected example
Econometrics and statistical techniques								
Difference in difference	Compares impacts between treatment and control (comparison) groups both before and after the implementation of an intervention.	×	×		×	×		
Regression analysis	Shows the degree of variation of samples around a linear or nonlinear relationship, and thus the statistical significance of the relationship.	×			×			The approach used to evaluate the impact of the Nicaraguan conditional cash transfer program, Red de Protección Social, on changes in household expenditure on food, improved health care, and the nutritional status of children (Moore, 2009).

(Continued)

Table 1.1. Continued

		Especially applicable to:			Addresses challenge of:			
Methodology	Definition	Simple	Complicated	Complex	Attribution	Baselines/Targets	Long time horizons	Selected example
Contextual monitoring	Monitors external trends or contexts.		×	×		×	×	In Nepal DFID uses qualitative data from a range of sources to monitor various indicators that are likely to influence program outcomes. This includes indicators on communal violence, human rights abuses, rule of law, and the role of representatives of marginalized groups in political institutions (DFID, 2012).
Sequential targeting	Sets interim targets or several milestones that relate to expected performance over short intervals and are revised over time.		×	×		×		DFID Violence Against Women and Girls (VAWG) programs aim to change social norms and ultimately prevent violence against women in a number of developing countries. However, long-term targets to achieve desired change in social norms are difficult to predict. VAWG evaluations therefore focus on sequential targeting to evaluate the long-term changes in social norms and reduced violence realistically (DFID, 2010).

Monitoring and evaluation tools

(Continued)

Table 1.1. Continued

Methodology	Definition	Especially applicable to:			Addresses challenge of:			Selected example
		Simple	Complicated	Complex	Attribution	Baselines/Targets	Long time horizons	
Contribution analysis	Focuses on how an intervention interacts with other aid or nonaid factors and analyzes whether an intervention was a necessary and/or a sufficient causal factor, along with other factors.			×	×	×		Broad thematic evaluations such as those carried out by the Organisation for Economic Co-operation and Development—Development Assistance Committee [OECD DAC] on donor gender mainstreaming (OECD, 2011) use a contribution approach.
Theories of change	An articulated theory of how the anticipated change will come about and the contribution to this of any activities.		×	×	×		×	A review for Care International has found that Theories of Change have a critical use in peace building. The review suggests that this approach helps deal with underlying assumptions, identifies aims and objectives, clarifies project design, and makes more explicit links between local level activities and national peace processes for desired changes to occur (Care International, 2012).

Monitoring and evaluation tools

a climate event, or there may not be an end state at which point the problems have been resolved (Rogers, 2008).

For simple interventions, where there is agreement and certainty, well-established monitoring and evaluation methodologies used in international development and other areas can be used to understand progress in the short to medium term. More complicated adaptation initiatives, where there is agreement but less certainty, may require a broader set of approaches (Rogers, 2008). However, many adaptation initiatives are complex, and the fundamental uncertainties associated with climate change create particular challenges for evaluation.

The challenges to consider in the M&E of adaptation of these types of interventions include attributing observed change to specific activities within complex contexts, setting baselines and targets with changing climatic hazards, and assessing the effectiveness of adaptation initiatives with long-term benefits within short- and medium-term evaluation cycles (Lamhauge, Lanzi, & Agrawala, 2012). Theoretical frameworks for monitoring and evaluating adaptation that have emerged in recent years at the project, program, and fund levels (Bours, McGinn, & Pringle, 2014; Brooks et al., 2013; Olivier, Leiter, & Linke, 2013; Pringle, 2011; Spearman & McGray 2011; Villanueva, 2011) provide overarching guidelines within which adaptation M&E can be conducted. However, adaptation planners and practitioners have struggled to find the methods within those guideline to address the challenge of combining accountability to donors, measuring results within the program or funding cycles, and considering the more particular challenges of assessing the effectiveness of climate change adaptation over the long term. This article therefore focuses on the appropriate methods that may be of use within these broader frameworks and an overarching approach to improve CCAR M&E.

Monitoring and Evaluation Approaches and Techniques Useful to Adaptation

Table 1.1 summarizes the evaluation methods reviewed for this article to address challenges in simple, complicated, and complex adaptation, including issues of attribution, shifting baselines and targets, and evaluation over long time horizons. The table shows what type of adaptation the methodology can address as well as which challenges it deals with.

We next discuss how the methods presented in the table can address each challenge for adaptation M&E, and the lessons learned from their application in other fields that are relevant to CCAR.

Addressing Attribution or Contribution

Demonstrating attribution can be challenging when interventions are complex or when the adaptation component is a relatively small part of a larger development program. This has been a long-standing challenge in international development and there are a variety of methods shown in Table 1.1 that can provide evidence on this issue, including social science methods such as focus groups and semi-structured interviews, statistical techniques and experimental or quasi-experimental research design and analysis. From the review of these methods and how they have been applied, we can draw several lessons for the evaluation of CCAR. First, when counterfactuals are not available, they can in some cases be inferred through quasi-experimental or participatory methods, but there will also be complex contexts where the nature of climate change is not well understood and it is difficult to establish a counterfactual. This is also the case in other instances of environmental policy (Ferraro, 2009). In complex contexts, it may be more meaningful to examine the contribution of an intervention to the observed outcome rather than to look for a direct causal attribution (Kotvjos & Shrimpton, 2007; Mayne, 2001). Second, to understand how to address attribution or contribution we must understand the mechanism that leads to the outcome—that is, the theory of change. Through theories of change, assumptions are made clear from the beginning, and mid-term goals can feed into long-term goals and thereby make the measurement or assessment of these pathways easier (Weiss, 1995).

Baselines and Target Setting

Different types of baseline and target-setting techniques are used in international development to address unknown future trends in societies and ecosystems (Dinshaw et al., 2014). To address the challenges of climate adaptation, these techniques need to take into account the shifting baselines of climate hazards and therefore moving targets for what is defined as "success." This can be particularly challenging, as the data on past or future climate trends may not be available, and future climate change is surrounded with uncertainty. As shown in Table 1.1, methods used in international development to address this challenge include surveys, various techniques for adjusting baselines and targets over time to a variety of futures, normalizing results for changes in underlying trends and contextualizing them with a narrative. The review in Dinshaw et al. (2014) shows that there are a range of existing techniques, including normalization and contextualization, that can contend with the shifting baselines specifically in adaptation projects. There are also techniques to monitor the changing external context and the interaction with project outcomes (Department for International Development, 2010; Hargreaves, 2010). When external factors change, indicators may need to be modified in order to reflect this change and sequential targeting may help define gradually evolving objectives. When necessary, baselines can be reconstructed and continually updated (Gakhar, Kaur, & Kapur, 2010; Pradhan & Rawlings, 2002; World Bank, 2010). Although there is a variety of methods that can be used to assess attribution and baselines and targets for interventions that focus on current climate variability, long-term horizons remain a particular challenge for the M&E of interventions seeking to build longer-term resilience to future climatic threats.

Uncertainty and Long Time Horizons

Although many development interventions need to be implemented and monitored over a long time, for example, improving an education system or enabling the transition to a democratic political structure, it is possible to know what success looks like for these interventions. However, what success looks like for adaptation to a changing climate is less clear, because there is uncertainty about both non-climatic and climatic futures. Although all development and conservation interventions are subject to uncertainty about the context, such as political situations and funding, adaptation projects are also subject to uncertainty about how the climate will change, how rapidly, and whether the models to predict this change are reliable and available at appropriate scales. Given the long-term nature of climate change, adaptation interventions that are building resilience to an uncertain climate future need to be monitored over long periods of time. However, there is no definite point at which evaluators can determine that a system or community has fully adapted while the climate is still changing.

There are relatively few methods for contending with uncertainty and long time horizons for CCAR M&E. The relatively few examples of longitudinal evaluations available include broad thematic evaluations such as those carried out by the Organisation for Economic Co-operation and Development (OECD), Development Assistance Committee on donor gender mainstreaming (gender mainstreaming being a process of considering the needs of both men *and women* in all forms of policies, processes, and implementation to support the goal of achieving gender equality) (OECD, 2011), the effectiveness of budget support (OECD, 2012), and a study by the United Nations Development Programme (UNDP) on a decade of efforts on strengthening national capacities in disaster risk management and recovery (UNDP, 2010). Another example of long-term monitoring is when a development agency evaluates a whole scope of programs over their entire period of implementation, such as the Sida evaluation of its support to Vietnam, Laos, and Sri Lanka (McGillivray, Pankhurst, & Carpenter, 2012). However, this type of long-term evaluation is rare for specific projects, and there are few methods that help with monitoring and evaluating adaptation over long time horizons. One method that is used in biodiversity conservation that may be applicable to ecosystem-based adaptation is limiting factor analysis, wherein the evaluators preemptively identify a list of factors that could limit the effectiveness of the project in the future and consider how to adjust program design accordingly (Gullison & Hardner, 2009). A similar approach being tested and developed is the Theory of No Change, which came out of a meta-evaluation of climate change mitigation evaluations supported by a community of practice hosted by the Independent Evaluation Office of the Global Environment Facility (GEF IEO). This theory enables evaluators to hypothesize why certain causal links may be broken or why interventions are not working in given circumstances (Woerlen, 2013).

Employing an Overarching Approach To Improve Monitoring and Evaluation of Climate Change Adaptation

In order to contend with the challenges of evaluating adaptation projects described in this volume—creating baselines, establishing attribution, and monitoring over long time horizons—evaluators need to move beyond employing specific M&E methodologies to address particular questions. There are three overarching approaches to M&E that could make CCAR M&E more effective when used within a conceptual framework: using mixed methods, including or adopting participatory approaches, and incorporating learning into the ongoing monitoring and evolving design of an intervention. Using such approaches within an overall conceptual framework that links them together, and theorizes the links between the chosen methods and key elements of the climate challenge in question, will go some way to addressing the challenges identified in this article.

Mixed Methods

Different M&E methods have different strengths and weaknesses in addressing the core challenges for climate change adaptation. However, we can see from their application in other development projects that they can be used together to complement each other, and using mixed methods can address some of the challenges of complicated or complex interventions. When applying different methods, a careful assessment of the underlying assumptions is required to ensure that the evidence gained from each method is compatible.

For example, to monitor and evaluate an intervention that provides monetary incentives and training to farmers to increase their resilience to climate change through land use change, a portfolio of methodologies was used (Leeuw & Vaessen, 2009). The methods included quantitative methods (e.g., experiment-related methods, econometrics, and statistical analyses) that enumerated outcomes, such as income levels, agricultural productivity, or access to services. Complementary qualitative methods examined how change had come about with the use of surveys, focus-group interviews, and participatory techniques. Each methodology had its comparative advantage that, when brought together, provided a more complete picture of the intervention outcome (Leeuw & Vaessen, 2009).

Generally, evaluation designs that use different types of methods offer opportunities for triangulation and complementarity and thus a deeper understanding and conclusions about causality. Using mixed methods can help assess the different aspects of contribution of the intervention to overall adaptation success. However, mixed methods do not help evaluators much with the challenges of setting baselines or monitoring over long time horizons.

Participatory Approaches

Participatory monitoring and evaluation is particularly valuable in complex adaptation contexts to assess changes in attitudes and decision making, to examine the impact and effectiveness of interventions, to create baselines and comparison groups, and to build ownership or recommendations. Data gathered in a participatory manner are only one input into monitoring and evaluating these contexts, but offer a methodologically simple way of understanding impacts on livelihoods and household experiences that may be difficult to capture through a set of indicators. Participatory methods capture unintended consequences and impacts, and involving beneficiaries and a wide range of stakeholders also ensures that evaluations are grounded in local realities. For instance, Heltberg et al. (2012) conducted an extensive survey after the 2008 global food and financial crisis to provide a narrative from the perspective of those who were affected, and to provide feedback on a range of relief schemes that were implemented after the crisis.

Participatory approaches can be very helpful in ensuring that the contributions of various elements to the success of the intervention are well understood. If this practice is continued throughout an intervention, evaluators can create and test narratives of attribution and theories of change (Brooks et al., 2013). However, sustaining participation for the purpose of monitoring interventions over long time frames may be challenging, and also places demands and potential opportunity costs on participants who may not directly experience the benefits of the activities (Cooke & Kothari, 2001).

A Focus on Learning

What makes a method most appropriate to climate change adaptation is not necessarily its intrinsic qualities, as this is highly contextual depending on the objectives of the intervention, but instead how the method is applied. One way to address the uncertainty and lack of information that challenge CCAR M&E is by applying methods that emphasize iterative monitoring techniques and enable learning over time. Several of the techniques identified in Table 1.1 can be used iteratively and with a focus on incorporating learning into the ongoing evaluation. For instance, a single baseline can be developed at the beginning of an intervention and then used for the duration of the intervention. Alternatively, the evaluators can make use of rolling baselines that are developed during different stages of the program, or stochastic baselines that are created for several different scenarios that vary in their underlying assumptions. These iterative techniques enable more flexibility to include learning and adjustment throughout the intervention.

Possible limitations of these methods include the capacity required to generate multiple baselines, some of which could include complex computational modeling, and the ability to interpret the new information generated continuously (Blanco-Fonseca, 2010). Furthermore, the application of rolling baselines may cause a bias where the program inputs and the outputs or outcomes become correlated with other factors that cannot be controlled or predicted. In this case, normalization techniques can enable the evaluator to separate and quantify the impact of different influencing factors on the final outcome (Pradhan & Rawlings, 2002).

Developmental evaluation is an existing overarching approach to address complex evaluation problems and contexts, which is suitable for CCAR M&E. It is a form of in-built, iterative evaluation that aims to help program staff use evaluation findings throughout the program cycle and in the ongoing refinement of the program (Patton, 2011). Developmental evaluation has a focus on systems and problems that are complex, characterized by uncertainty and dynamic change. It does not refer to a specific set of methods or tools but rather a mindset of inquiry, and is thereby focused on bringing data and lessons from the evaluation process into the ongoing program through careful timing and stakeholder participation. Rather than performing evaluations based on a predetermined schedule, developmental evaluations seek to coincide with annual workflow plans, major implementation steps, and decisions about the future of programs (Patton, 2011).

Conclusion

Adaptation M&E can and should learn from the M&E experiences in other fields. There is a wide range of experience and tested methods that can help address issues of attributing impacts and some aspects of uncertain futures. However, it seems clear that the particular issues of uncertainty about climate change and the long time horizons over which adaptation interventions need to be implemented can make the monitoring and evaluation of complex interventions in this context particularly challenging. Therefore, we must move beyond employing specific methods and techniques in isolation, but select and combine those that together provide sensitivity to the challenges of uncertainty and longer time frames. As the experience of M&E in the field of climate change adaptation evolves, it will offer important lessons to those carrying out evaluations of complex interventions in other areas and into distant futures of high uncertainty.

References

Adaptation Committee. (2014). *Report on the workshop on the monitoring and evaluation of adaptation from the fifth meeting of the Adaptation Committee.* Bonn, Germany:

UNFCCC Secretariat. Retrieved from http://unfccc.int/files/adaptation/cancun_
adaptation_framework/adaptation_committee/application/pdf/ac_me_ws_report_
final.pdf

American Dietetic Association. (2008). Nutrition care process and model part I:
The 2008 update. *Journal of the America Dietetic Association, 108*, 1113–1117. doi:
10.1016/j.jada.2008.04.027

Blanco-Fonseca, M. (2010). *Literature review of methodologies to generate baselines for
agriculture and land use*. Common Agricultural Policy Regional Impact—The Rural
Development Dimension (CAPRI-RD). Bonn, Germany: Universität Bonn. Retrieved
from http://www.ilr.uni-bonn.de/agpo/rsrch/capri-rd/docs/d4.1.pdf

Bours, D., McGinn, C., & Pringle, P. (2014). *Monitoring and evaluation for climate change
adaptation and resilience: A synthesis of tools, frameworks and approaches* (2nd ed.)
Phnom Penh, Cambodia: SEA Change CoP, and Oxford, United Kingdom: UKCIP.
Retrieved from http://www.ukcip.org.uk/wordpress/wp-content/PDFs/SEA-Change-
UKCIP-MandE-review-2nd-edition.pdf

Brooks, N., Anderson, S., Burton, I., Fisher, S., Rai, N., & Tellam, I. (2013). *An operational framework for tracking adaptation and measuring development* (Climate Change
Working Paper No. 5). London, United Kingdom: International Institute for Environmental Development. Retrieved from http://pubs.iied.org/pdfs/10038IIED.pdf

Buchner, B., Stadelmann, M., Wilkinson, J., Mazza, F., Rosenberg, A., & Abramskiehn,
D. (2014). *The global landscape of climate finance 2014*. San Francisco, CA: Climate
Policy Initiative. Retrieved from http://climatepolicyinitiative.org/publication/global-
landscape-of-climate-finance-2014/

Care International. (2012). *Guidance for designing, monitoring and evaluating peace-
building projects: Using theories of change*. London, United Kingdom: Care International. Retrieved from http://conflict.care2share.wikispaces.net/file/view/CARE+
International+DME+for+Peacebuilding.pdf

Cooke, B., & Kothari, U. (2001). *Participation: The new tyranny?* London, United Kingdom: Zed Books.

Davies, R., and Dart, J. (2005). *The "Most Significant Change" (MSC) technique: A guide to
its use*. Melbourne, Australia: MandE. Retrieved from http://www.mande.co.uk/docs/
MSCGuide.pdf

Department for International Development. (2010). *Working effectively in conflict-
affected and fragile situations*. Monitoring and evaluation briefing paper 1. London,
United Kingdom: Department for International Development.

Department for International Development. (2012). *Violence against women and girls—
Guidance on monitoring and evaluation for programming on violence against women and
girls* (Guidance Note 3 in CHASE Guidance Note Series). London, United Kingdom:
Department for International Development.

Dinshaw, A., Fisher, S., McGray, H., Rai, N., & Schaar, J. (2014). *Monitoring and eval-
uation of climate change adaptation: Methodological approaches* (OECD Environment
Working Papers No. 74). Paris, France: Organisation for Economic Co-operation and
Development Publishing. Retrieved from http://dx.doi.org/10.1787/5jxrclr0ntjd-en

Ferraro, P. J. (2009). Counterfactual thinking and impact evaluation in environmental
policy. In M. Birnbaum & P. Mickwitz (Eds.), Environmental program and policy
evaluation: Addressing methodological challenges. *New Directions for Evaluation, 122*,
75–84. doi:10.1002/ev.297

Gakhar, K. D., N. Kaur, & V. Kapur. (2010). *Propensity score matching method in quasi-
experimental designs: An approach to programme evaluation of INHP-III* (Discussion Paper 3). New Delhi, India: Sambodhi Research & Communications Pvt. Ltd.

Gullison, R., & Hardner, J. (2009). Using limiting factors analysis to overcome the
problem of long time horizons. Environmental program and policy evaluation: Addressing methodological challenges. *New Directions for Evaluation, 122*, 19–29. doi:
10.1002/ev.292

Hargreaves, M. (2010). *Evaluating system change: A planning guide*. Princeton, NJ: Mathematica Policy Research, Inc.

Heltberg, R., Hossain, N., & Reva, A. (Eds.). (2012). *Living through crises. How the food, fuel and financial shocks affect the poor*. Washington, DC: World Bank. Retrieved from https://openknowledge.worldbank.org/handle/10986/6013

Hyse, H., & Ongevalle, J. (2008). *Fulfilling expectations? The experiences with the M&E-part of outcome mapping in an education for sustainability project in Zimbabwe*. Leuven, Belgium: Katholieke Universiteit Leuven.

Independent Evaluation Group. (2012). *Adapting to climate change: Assessing the World Bank Group experience phase III*. Washington, DC: Independent Evaluation Group, World Bank. Retrieved from http://ieg.worldbankgroup.org/Data/reports/cc3_full_eval_0.pdf

India HIV/Aids Alliance. (2007). *Stories of significance: Redefining change. An assortment of community voices and articulations*. Washington, DC: World Bank. Retrieved from http://gametlibrary.worldbank.org/FILES/996_MSC%20Report%20on%20HIV%20and%20AIDS%20in%20India.pdf

Klein, L., Clayton, S., Alldredge, R., & Goodwin, P. (2007). Long-term monitoring andevaluation of the Lower Red River meadow restoration project, Idaho, U.S.A. *Restoration Ecology, 15*(2), 223–239. doi: 10.1111/j.1526–100X.2007.00206.x

Kotvojs, F., & Shrimpton, B. (2007). Contribution analysis: A new approach to evaluation in international development. *Evaluation Journal of Australasia, 7*(1), 27–35.

Lamhauge, N., Lanzi, E., & Agrawala, S. (2012). *Monitoring and evaluation for adaptation: Lessons from development co-operation agencies* (OECD Environment Working Paper No. 38). Paris, France: Organisation for Economic Co-operation and Development Publishing. doi:10.1787/19970900

Leeuw, F., & Vaessen, F. (2009), *Impact evaluations and development. NONIE guidance on impact evaluation*. Washington, DC: Independent Evaluation Group, World Bank. Retrieved from http://siteresources.worldbank.org/EXTOED/Resources/nonie_guidance.pdf

Mayne, J. (2001). Addressing attribution through contribution analysis: Using performance measures sensibly. *The Canadian Journal of Program Evaluation, 16*(1), 1–24.

McGillivray, M., Pankhurst, A., & Carpenter, D. (2012). *Long-term Swedish development cooperation with Sri Lanka, Vietnam and Laos. Synthesis of evaluation study findings*. Stockholm, Sweden: Swedish International Development Corporation (Sida).

Moore, C. (2009). *Nicaragua's red de protección social: An exemplary but short lived cash transfer programme*. Brasilia, Brazil: International Policy Centre for Inclusive Growth. Retrieved from http://hdl.handle.net/10419/71770

Olivier, J., Leiter, T., & Linke, J. (2013). *Adaptation made to measure: A guidebook to the design and results-based monitoring of climate change adaptation projects* [Manual] (2nd ed.). Eschborn, Germany: Deutsche Gesellschaft für Internationale Zusammenarbeit (GIZ) GmbH. Retrieved from https://gc21.giz.de/ibt/var/app/wp342deP/1443/wp-content/uploads/filebase/me/me-guides-manuals-reports/GIZ-2013_Adaptation_made_to_measure_second_edition.pdf

Organisation for Economic Co-operation and Development. (2011). *Mainstreaming gender equality. Emerging evaluation lessons* (Evaluation insights no. 3). Paris, France: Author. Retrieved from http://www.oecd.org/derec/afdb/50313658.pdf

Organisation for Economic Co-operation and Development. (2012). *Evaluating budget support. Methodological approach*. Paris, France: Author. Retrieved from http://www.oecd.org/dac/evaluation/dcdndep/Methodological%20approach%20BS%20evaluations%20Sept%202012%20_with%20cover%20Thi.pdf

Patton, M. Q. (2011). *Developmental evaluation. Applying complexity concepts to enhance innovation and use*. New York, NY: The Guilford Press.

Pradhan, M., & Rawlings, L. (2002). The impact and targeting of social infrastructure investments: Lessons from the Nicaraguan Social Fund. In World Bank Economic Review (Vol. 16, pp. 275–295). Washington, DC: World Bank.

Pringle, P. (2011). AdaptME toolkit: Adaptation monitoring and evaluation. Oxford, United Kingdom: UK Climate Impacts Programme (UKCIP). Retrieved from http://www.ukcip.org.uk/wordpress/wp-content/PDFs/UKCIP-AdaptME.pdf

Rogers, P. (2008). Using programme theory to evaluate complicated and complex aspects of interventions. Evaluation, 14, 29–48. doi:10.1177/1356389007084674

Spearman, M., & McGray, H. (2011). Making adaptation count: Concepts and options for monitoring and evaluation of climate change adaptation. Eschborn, Germany: Gesellschaft für Internationale Zusammenarbeit (GIZ) GmbH. Retrieved from http://pdf.wri.org/making_adaptation_count.pdf

United Nations Development Programme. (2010). Evaluation of UNDP contribution to disaster prevention and recovery. New York, NY: Author. Retrieved from http://erc.undp.org/evaluationadmin/downloaddocument.html?docid=4397

U.S. Agency for International Development. (2012). Feed the Future, US government's global hunger and food security initiative [M&E guidance series]. Washington, DC: United States Agency for International Development.

Villanueva, P. S. (2011). Learning to ADAPT: Monitoring and evaluation approaches in climate change adaptation and disaster risk reduction—Challenges, gaps and ways forward (SCR Discussion Paper 9). Brighton, United Kingdom: Institute of Development Studies. Retrieved from http://www.ids.ac.uk/files/dmfile/SilvaVillanueva_2012_Learning-to-ADAPTDP92.pdf

Weiss, C. (1995). New approaches to evaluating comprehensive community initiatives. Washington, DC: The Aspen Institute. Retrieved from http://files.eric.ed.gov/fulltext/ED383817.pdf

White, H., Sinha, S., & Flanagan, A. (n.d.). A review of the state of evaluation. Washington DC: Independent Evaluation Group, World Bank. Retrieved from http://www.oecd.org/development/evaluation/dcdndep/37634226.pdf

Woerlen, C. (2013). The theory of no change. Evaluation connections: The EES Newsletter. Retrieved from https://www.climate-eval.org/sites/default/files/news/ees%20newsletter.pdf

World Bank. (2010). Reconstructing baseline data for impact evaluation and results measurement. Part of a special series on nuts and bolts of M&E. Washington, DC: World Bank. Retrieved from http://siteresources.worldbank.org/INTPOVERTY/Resources/335642--1276521901256/premnoteME4.pdf

Zimmerman, B., & Glouberman, S. (2004). Complicated and complex systems: What would successful reform of Medicare look like? In P. G. Forest, T. McIntosh, & G. Marchildon (Eds.), Health care services and the process of change (pp. 21–53). Toronto, Canada: University of Toronto Press.

Zimmerman, B., Lindberg, C., & Plsek, P. (1998). Edgeware: Insights from complexity science for health care leaders. Irving, TX: VHA.

SUSANNAH FISHER is a senior researcher at the International Institute of Environment and Development (IIED), United Kingdom.

AYESHA DINSHAW is an associate at the World Resources Institute (WRI), United States of America.

NEW DIRECTIONS FOR EVALUATION • DOI: 10.1002/ev

HEATHER McGRAY is the Director of the Vulnerability and Adaptation Initiative, World Resources Institute (WRI), United States of America.

NEHA RAI is a senior researcher at the International Institute of Environment and Development (IIED), United Kingdom.

JOHAN SCHAAR is the Head of Development Cooperation in the Consulate General of Sweden, Jerusalem.

NEW DIRECTIONS FOR EVALUATION • DOI: 10.1002/ev

Krause, D., Schwab, M., & Birkmann, J. (2015). An actor-oriented and context-specific frame-work for evaluating climate change adaptation. In D. Bours, C. McGinn, & P. Pringle (Eds.), *Monitoring and evaluation of climate change adaptation: A review of the landscape. New Directions for Evaluation, 147,* 37–48.

2

An Actor-Oriented and Context-Specific Framework for Evaluating Climate Change Adaptation

Dunja Krause, Maria Schwab, Jörn Birkmann

Abstract

Responding to recently identified challenges in the comprehensive evaluation of climate change adaptation, we suggest an assessment framework that incorporates a multidisciplinary viewpoint on adaptation. The framework combines three major components to link system-oriented concepts of risk and vulnerability to natural hazards and climate change with actor-oriented approaches of decision-making: the risk context, individual decision-making, and adaptation assessment. We have thereby developed an approach that is embedded in context and scale, takes an actor-oriented perspective, and is applicable to multiple evaluation approaches. We tested the framework empirically with a mixed-methods approach to assess adaptation in flood-exposed areas of the Vietnamese Mekong Delta. This has shown the distinct meanings of good adaptation and uncovered mismatches in choices for adaptation strategies and evaluations of those strategies across different actors hindering sustainable development. © 2015 Wiley Periodicals, Inc., and the American Evaluation Association.

M onitoring and evaluation (M&E) for climate change adaptation (CCA) has emerged in various contexts and aims at identifying good adaptation, often by providing quantitative evidence for the quality of a strategy – for example via efficiency-based cost–benefit analyses (cf., e.g., Mechler, 2008; Stern, 2007; World Bank, 2010). Other CCA evaluation approaches highlight social aspects, and measure adaptation success based on vulnerability reduction or capacity building. Common to most is the assessment of adaptation against a predefined goal set as a preferable alternative to the status quo. For development-oriented CCA projects, input–output–outcome evaluations (IOOE) and multicriteria decision analyses (MCDA) rank prominently among evaluation schemes. IOOE commonly rely on "logframe"-based analyses that describe processes by assessing how inputs, activities, and outputs are causally linked and determine outcomes they have on a system (cf. Global Environment Facility Independent Evaluation Office [GEF IEO], 2007, p. 8; Schwab, 2014, p. 32). MCDAs are "methods and procedures by which concerns about multiple conflicting criteria can be formally incorporated in a decision-making process" (International Society on Multi-Criteria Decision-Making 2004, as cited in Gamper, Thöni, & Weck-Hannemann, 2006, p. 294). Most approaches focus on the implications of a strategy for a specific target group (cf. Ayers, Anderson, Pradhan, & Rossing, 2012; Oxfam, 2008), and/or account for specific spatial and temporal scales (Centre for International Studies and Cooperation [CECI], 2010; Junghans & Harmeling, 2012).

It is often neglected that actions are part of complex systems and change their meaning and value in relation to other strategies and actors (Adger, Lorenzoni, & O'Brien, 2009; Moser & Ekstrom, 2010). Good adaptation for one person can be bad for another; so that individual goals and values are important to consider. It is debatable to what extent good can be measured in absolute terms with a one-dimensional approach (Grothmann & Reusswig, 2006; O'Brien & Wolf, 2010; Schwab, 2014).

This chapter proposes an assessment framework that focuses on subjective perspectives on adaptation (i.e., perceptions and judgments of individual decision-makers) and links them with objective perspectives. The framework integrates a variety of criteria, assesses implications for different actors, and goes beyond a single scale. It can therefore identify mismatches between different values, goals, and differential judgments that result from the chosen evaluation approach. The endeavor of strategically linking actor-oriented approaches to adaptation with system-oriented vulnerability

At the time of research, the authors were affiliated with the United Nations University Institute for Environment and Human Security (UNU-EHS). This research was carried out at UNU-EHS and funded by the German Ministry of Education and Research (BMBF) and is based on the doctoral dissertations of Dunja Krause and Maria Schwab. Correspondence concerning this article should be addressed to krause@unrisd.org.

concepts is at the center of our analysis in an attempt to provide a new, more comprehensive lens for CCA evaluation.

In the following sections, we introduce our empirically tested conceptual framework before discussing the comparative advantages of the approach and its potential for facilitating promising adaptation strategies.

Conceptual Framework

The suggested framework combines three major analytical components, each linked to different schools of thought: the risk context, individual decision-making, and adaptation assessment (Figure 2.1, based on ideas of Grothmann & Patt, 2005; United Nations Framework Convention on Climate Change [UNFCCC], 2010; Werlen, 1993). We argue that decision-making about climate change adaptation, and the perceived quality of climate change adaptation interventions, are highly dependent on individual risk appraisal, and embedded into the specific risk context at place. The assessment of adaptation alternatives should therefore comprise an appraisal of both the risk and decision-making contexts, and a critical (self-) assessment of the evaluation approach applied.

Figure 2.1. Conceptual Framework for Adaptation Evaluation

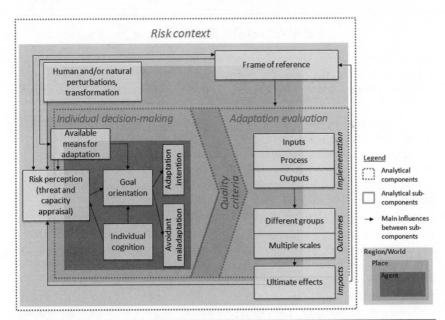

The risk context provides a basis for assessing the vulnerability-related outcomes of a strategy against the background of a given baseline and determines quality criteria for multi-criteria analyses. Considering decision-making processes does not only allow accounting for subjective goals and value systems for different actor groups but links multi-criteria-based quality judgment to decisions for or against certain measures. In combination with an assessment of the actual adaptation process, as part of a logframe-based approach, these components enable a more comprehensive approach for meeting challenges of CCA evaluation.

Risk Context

Risk is conceptualized as a function of hazard, meaning an extreme natural event or process, and vulnerability of the hazard-exposed system. Following the work of O'Brien, Eriksen, Schjolden, and Nygaard (2004), vulnerability is defined as system-inherent characteristic that shapes the extent of damage a hazard can bring about. It is a function of a system's exposure, susceptibility, and capacity of response (cf., e.g., Birkmann et al., 2013; Turner et al., 2003). We build on a social–ecological understanding of risks of natural hazards, based on the understanding that complex, cross-scale interactions between social and ecological systems result in risky environments. Such risks (composed of environmental hazards and social vulnerability) and the perception of them influence adaptation goals and intentions that are not necessarily limited to a specific hazard but dependent on the broader livelihood situation of the respective actor.

This conceptualization of risk implies generic and hazard-specific vulnerability factors against which one can compare the outcomes of adaptation. Given that adaptation as a process and adaptation outcomes can only be understood when accounting for their dynamics, it is important to consider feedback effects and transformations of the risk context that may result from social–ecological interactions.

Individual Decision-Making

Any action is the outcome of a decision based on perceptions and socio-cognitive drivers. We therefore build on concepts of adaptation appraisals from the climate change and disaster-risk community (e.g., Grothmann & Patt, 2005; Wisner, Blaikie, Cannon, & Davies, 2004) and social psychology (e.g., Lazarus & Folkman, 1984). Understanding *risk perception* is required for understanding decisions for or against adaptation to climate change. Corresponding to the important distinction of primary and secondary appraisal made by Lazarus & Folkman (1984), it comprises both the perception of threats, meaning potential negative impacts a hazard may provoke, and the perception of capacities of response, which may give an actor the feeling of being either empowered or helpless (cf. Grothmann & Reusswig, 2006). The perceived capacities include an individual appraisal of the expected outcomes of a strategy and the required means to implement this strategy. For adaptation to take place, actors therefore need not only to perceive climate change as a threat, but also judge an option positively with respect to efficacy and rate their own capacities sufficient to achieve a positive adaptation outcome. Otherwise, "avoidant maladaptation," the evasion of action, may arise, resulting from a feeling of incapability to respond adequately (Grothmann & Patt, 2005; Grothmann & Reusswig, 2006). Given that such decisions are critically linked to individual goals and priorities,

NEW DIRECTIONS FOR EVALUATION • DOI: 10.1002/ev

long-term CCA often competes with more pressing livelihood concerns and loses priority in everyday decision-making (cf. Intergovernmental Panel on Climate Change [IPCC], 2012, p. 45).

Assessing adaptation prospects thus must consider the various, potentially conflicting goals pursued by the adapting agent(s). It is crucial to identify possible conflicts on at least two levels: at the level of the actor who chooses and prioritizes among goals based on the specific situation and at the level of social interaction, as different strategies are often negotiated and decided upon by multiple actors with differing levels of power (cf. Edvardsson Björnberg & Svenfeldt, 2009).

Adaptation Assessment

To understand what adaptation looks like, we build on logframes to assess the process of adaptation and account for adaptation outcomes and impacts (Global Environment Facility Independent Evaluation Office [GEF IEO], 2007). This comprises a description of inputs in the form of time, financial, natural, and physical assets or capital needed to implement adaptation. The characteristics of the actual adaptation action describe how these inputs are used, exchanged, and transformed (Jacob & Mehiriz, 2012). This encompasses institutional or governance criteria essential for process-based evaluations (Jacob & Mehiriz, 2012). Immediate products and services define outputs; more short- and medium-term effects represent outcomes; pervasive and longer-term results of adaptation are impacts (GEF IEO, 2007, p. 5; United Nations Framework Convention on Climate Change [UNFCCC], 2010, p. 5).

The value of adaptation and individual decisions is more complex than that can be measured by common aspects included in a logframe, however. We integrated individual actors' goals and quality criteria into the adaptation assessment to pinpoint diverging appraisals of the best alternative. Thus, we combine aspects of logframe approaches and multicriteria decision analysis (MCDA) (cf. Eakin & Bojórquez-Tapia, 2008; Linkov et al., 2006). MCDAs provide a subjective perspective to adaptation assessment as they consider subjective weightings and judgments of the various actors involved.

Addressing the risk context serves as baseline for a vulnerability centered definition of successful strategies. This identifies criteria most relevant to the actors involved in terms of both process of implementation and expected outcome of adaptation. The decision-making perspective explains why and how adaptation intentions are formed and supports a better understanding of differential motivations to act and of diverse priorities negotiated among different actors.

Methodologically, this approach poses several challenges. It requires data for the strategic analysis of multidimensional processes, subjective and objective perspectives, and complex cross-scale interactions. We therefore apply a mixed-methods approach, based on a purposefully selected set of qualitative and quantitative methods in combination with a deliberate and flexible procedure that ensured participation of the local population in the design and application of the evaluation.

Assessing Local Adaptation to Flood Risks in the Vietnamese Mekong Delta

The Vietnamese Mekong Delta (VMD) is often ranked as one of the regions most at risk of climate change presently and particularly in the future, as it is exposed to sea-level rise which reinforces water-related hazards already threatening the population (cf. Ministry of Natural Resources and Environment, 2012). Our study focused on different adaptation interventions undertaken in response to flooding. At the government level, several CCA programs such as the National Target Program to Respond to Climate Change (Government of Vietnam, 2008) have been implemented and include good governance requirements. Yet participation and fairness as guiding principles are still neglected (Fortier, 2010). To counteract this, we included local voices of adaptation in our analysis and put them at the center of our assessment, which includes six steps:

1. Understand risk *context.*
2. Identify existing and potential adaptation *strategies* (different levels and stakeholders).
3. Identify *criteria* that determine success/failure of strategies and that need to be considered to ensure sustainability.
4. Identify *stakeholder preferences* for adaptation according to criteria weightings.
5. *Evaluate* adaptation strategies considering institutional setup and influence of different stakeholders.
6. Compare the results of the multicriteria-based evaluation results with vulnerability-related outcomes of adaptation strategies.

The use of this six-step assessment method has shown that meanings of "good" adaptation are varied and inconsistent. Goals of adaptation vary, sometimes substantially, across actors. Agricultural communities, for instance, placed greater emphasis on a strategy's monetary aspects, and households in urban areas emphasized impacts more in terms of opportunity costs, trade-offs, and quality of life. Some evaluation criteria such as environmental sustainability were important to authorities but not to households. These preferences influenced different actors' overall judgment on adaptation. In the urban element of the study, exposure reduction (by moving away) ranked comparatively low in households' priority setting

Figure 2.2. Weighting of Decision-Making Criteria for the Selection of Flood Adaptation Strategies

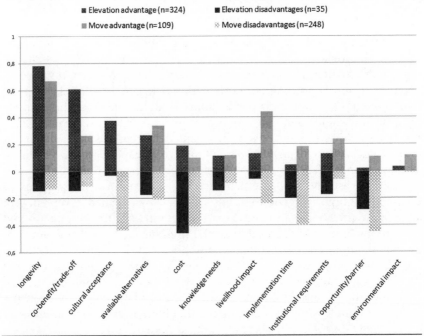

The figure highlights the key advantages and disadvantages household survey respondents in the urban case study saw in the cases of a generally favored option (house elevation in the flood-exposed area, judged as good strategy by 89.7 % of households) and a much more contested strategy (moving out of the flood zone, judged negatively by 70.4% of households).

Note: "Figure 7.2: Weighting of Decision-Making Criteria for the Selection of Flood Adaptation Strategies" by Dunja Krause (Krause, 2015, p. 166). Copyright 2015 by Dunja Krause.

(cf. Figure 2.2, from Krause, 2015). This conflicted with governmental goals of urban upgrading and resettlement.

An analysis of vulnerability implications across temporal scales of various adaptation strategies revealed that some strategies seemed promising at present but would almost certainly increase vulnerability in the long run: Most urban households, for example, favored the elevation of houses over moving to less flood-exposed areas as they perceived elevation to reduce flood impacts fast and at lower cost. Given the increasing exposure to flooding, opting for house elevation renders households more vulnerable in the long run, however, and the recurrent investments necessary because of continuously rising flood levels tie up limited financial resources, so that moving out of exposure becomes less achievable.

This comparison of vulnerability-oriented and outcome-based assessments with a multicriteria decision analysis pointed at diverging quality judgments and mismatches between preferences and objectively assessed implications.

Discussion

The VMD study did not aim at an absolute ranking of adaptation strategies for the Mekong Delta. Instead, it provided a basis for understanding the differential meaning of "good" adaptation across actors, regions, and time scales. The strong actor orientation risks overemphasizing aspects that may not be in line with sustainability goals, as shown at the example of most households' ignorance of environmental criteria (cf. Figure 2.2). We nevertheless opted for this actor-oriented approach as it can point out *mismatches between different values and goals* that shape adaptation realities. The dominant preference of most government representatives for exposure-reducing measures (e.g., resettlement) and opposition of households illustrated highly divergent risk appraisals and associated divergence of adaptation preferences. Incorporating diverging quality judgments and evaluations based on technical expertise presents a key challenge for program evaluation. The evaluation of many large-scale measures, for instance, resettlement programs, is mostly informed by expert judgments, whereas conflicting appraisals of affected households are rarely considered. Such an evaluation-based mismatch raises the question of whose judgment and what measure to opt for—a long-term oriented vulnerability reduction measure or a strategy that is locally accepted and supported.

To find a compromise requires awareness of these pitfalls and entails a negotiation based on goals, values, and culture. *Culturally competent evaluation* is a prerequisite for sustainable solutions. SenGupta, Hopson, and Thompson-Robinson (2014) argued that "[c]ulture shapes values, beliefs and worldviews. Evaluation is fundamentally an endeavor of determining values, merits, and worth" (p. 6). In the case of Vietnam, the technocratic expertocracy (e.g., Ehlert, 2012), which selectively consults a specific type of expert, favors the values of technocrats and engineers over those of the population vulnerable to climate change impacts. Although there might be justifiable reasons behind this preference, it comes with blind spots that may cause unsustainable development. Leaving out culturally sensitive criteria which shape people's perceptions and goals ignores the fact that people are constantly working their way around official approaches, undermining their success and investing more resources than necessary. Culturally sensitive approaches could increase acceptance of official measures and contribute to more inclusive and sustainable adaptation.

This context-specific and actor-oriented approach facilitates a strategic identification and diagnostic analysis of given adaptation barriers as demanded for by various authors (e.g., Dow et al., 2013; Moser & Ekstrom, 2010). A consideration of the risk context facilitated identifying barriers at the level of societal, economic, and environmental determinants and indicating strategic ways of overcoming them by, for instance, moving out of a hazard-exposed area. Actor orientation identified cognitive and knowledge-related obstacles (e.g., misperceptions regarding threats and capacities) and uncovered goal conflicts between different actors. Overcoming such barriers requires, among other things, an appreciation of each other's goals more than an effective adaptation strategy. The inclusion of the local population in the planning process can stimulate dialogue fostering such appreciation. A program-evaluation perspective added a quality dimension to adaptation appraisal and revealed that participatory approaches are valuable in judging the quality of different adaptation options. In the Vietnamese context, effective stakeholder participation in program evaluation could be implemented as part of the existing data collection, if a system of independent evaluators was established.

Conclusion

A critical reflection of evaluation schemes applied in a CCA context is essential for capturing the complexity of factors that influence adaptation and its value for different actors. The presented framework provides a solid foundation for a context-specific and actor-oriented assessment but also holds significant relevance for program evaluation of adaptation. The assessment goes beyond countable variables and considers factors important to local actors, making it particularly interesting for transdisciplinary evaluation research and practice. This facilitates more informed and integrative decision-making. The approach also illustrates significant challenges to evaluation endeavors in general, asking whose values count if a choice has to be made among conflicting options. It underscores that the intrinsic goal of evaluation determines its outcome. This challenges the increasing demand of donors for standardized evaluation approaches, instead calling for more deliberate and flexible designs of M&E program components.

The suggested approach recedes from absolute judgments and promotes good adaptation as a more informed and preference-based negotiation process. It can include nonevents as it addresses people's goals and preferences independent of climate change-related hazards resulting in a profound understanding of the risk context and potential adaptation barriers. The focus on perceptions and decision-making processes supports flexible but robust climate change adaptation.

In conclusion, our approach has *not* identified ultimate adaptations. On the contrary, we suggest receding from absolute judgments in CCA M&E. Even if achieved, it is debatable to what extent such evaluation results are reliable and valid across scales. The suggested approach instead emphasizes the distinct nature and diverse values of adaptation, calling for an inclusive and transparent negotiation process in the evaluation of CCA programs and actions, prioritizing flexible and adaptive measures.

References

Adger, N. W., Lorenzoni, I., & O'Brien, K. L. (Eds.). (2009). *Adapting to climate change: Thresholds, values, governance.* Cambridge, NY: Cambridge University Press. Retrieved from http://dx.doi.org/10.1111/j.1475–4959.2010.00360_5.x

Ayers, J., Anderson, S., Pradhan, S., & Rossing, T. (2012). *Participatory monitoring, evaluation, reflection and learning for community-based adaptation: A manual for local practitioners.* London, United Kingdom: Care International. Retrieved from http://www.careclimatechange.org/files/adaptation/CARE_PMERL_Manual_2012.pdf

Birkmann, J., Cardona, O. D., Carreño, M. L., Barbat, A. H., Pelling, M., Schneiderbauer, S., & Welle, T. (2013). Framing vulnerability, risk and societal responses: The MOVE framework. *Natural Hazards, 67*(2), 193–211. Retrieved from http://link.springer.com/article/10.1007%2Fs11069-013-0558-5

Centre for International Studies and Cooperation (CECI). (2010). *Framework on community based disaster risk management in Vietnam.* Montréal, Canada: CECI. Retrieved from http://www.ceci.ca/assets/Asia/Asia-Publications/CBDRM-Framework.pdf

Dow, K., Berkhout, F., Preston, B. L., Klein, R., Midgley, G., & Shaw, M. R. (2013). Limits to adaptation. *Nature Climate Change, 3*(4), 305–307. doi:10.1038/nclimate1847

Eakin, H., & Bojórquez-Tapia, L. (2008). Insights into the composition of household vulnerability from multi-criteria decision analysis. *Global Environmental Change, 18*(1), 112–127. doi:10.1016/j.gloenvcha.2007.09.001

Edvardsson Björnberg, K., & Svenfeldt, A. (2009). *Goal conflicts in adaptation to climate change: An inventory of goal conflicts in the Swedish sectors of the built environment, tourism and outdoor recreation, and human health* (Base data report). Stockholm, Sweden: Swedish Defence Research Agency (FOI). Retrieved from http://www.foi.se/Global/Kunder%20och%20Partners/Projekt/Climatools/Rapporter%20och%20artiklar/Goal_conflicts_adaption_climate_change.pdf

Ehlert, J. (2012). *Beautiful floods: Environmental knowledge and agrarian change in the Mekong Delta, Vietnam.* Münster, Germany: ZEF Development Studies Lit Verlag.

Fortier, F. (2010). Taking a climate chance: A procedural critique of Vietnam's climate change strategy. *Asia Pacific Viewpoint, 51*(3), 229–247. doi:10.1111/j.1467-8373.2010.01428.x

Gamper, C. D., Thöni, M., & Weck-Hannemann, H. (2006). A conceptual approach to the use of cost benefit and multi criteria analysis in natural hazard management. *Natural Hazards and Earth System Science, 6,* 293–302. doi:10.5194/nhess-6-293-2006

Global Environment Facility Independent Evaluation Office (GEF IEO). (2007). *GEF impact evaluation: Final report on a proposed approach to GEF impact evaluation.* Impact evaluation information document no. 2. Washington, DC: Author. Retrieved from https://www.thegef.org/gef/sites/thegef.org/files/documents/Impact_Eval_Infodoc2.pdf

Government of Vietnam. (2008). *National target program to respond to climate change* (Decision no. 158/2008/QD-TTg). Hanoi, Vietnam: Author.

Grothmann, T., & Patt, A. (2005). Adaptive capacity and human cognition: The process of individual adaptation to climate change. *Global Environmental Change, 15*(3), 199–213. doi:10.1016/j.gloenvcha.2005.01.002

Grothmann, T., & Reusswig, F. (2006). People at risk of flooding: Why some residents take precautionary action while others do not. *Natural Hazards, 38*(1–2), 101–120. doi:10.1007/s11069-005-8604-6

Intergovernmental Panel on Climate Change (IPCC). (2012). *Managing the risks of extreme events and disasters to advance climate change adaptation* [Special Report of the IPCC]. New York, NY: Cambridge University Press. Retrieved from http://www.ipcc-wg2.gov/SREX/images/uploads/SREX-All_FINAL.pdf

Jacob, J., & Mehiriz, K. (2012). *Elements of a frame of reference for evaluating adaptation to climate change: The RAC-Québec case* [Research report]. Québec, Canada: Centre de

Recherche et d'Expertise en Évaluation (CREXE). Retrieved from http://www.crexe .enap.ca/CREXE/Publications/Lists/Publications/Attachments/66/CREXE%20Researc h%20report%20(English)%20-%20Elements%20of%20a%20frame%20of%20referenc e_%C3%A9lectro.pdf

Junghans, L., & Harmeling, S. (2012). *Different tales from different countries: A first as- sessment of the OECD "Adaptation Marker"* [Briefing paper]. Bonn, Germany: German- watch. Retrieved from http://germanwatch.org/fr/download/7083.pdf

Krause, D. (2015). *Arenas of adaptation. A stakeholder-based evaluation of selected climate change adaptation strategies in Can Tho City, Vietnam* (Doctoral dissertation, Univer- sität Bonn, Bonn, Germany).

Lazarus, R. S., & Folkman, S. (1984). *Stress, appraisal, and coping.* New York, NY: Springer Publishing.

Linkov, I., Satterstrom, F. K., Kiker, G., Batchelor, C., Bridges, T., & Ferguson, E. (2006). From comparative risk assessment to multi-criteria decision analysis and adaptive management: Recent developments and applications. *Environment Interna- tional, 32*(8), 1072–1093. doi:10.1016/j.envint.2006.06.013

Mechler, R. (2008). *The cost–benefit analysis methodology* (Risk to Resilience Work- ing Paper No. 1). Retrieved from http://i-s-e-t.org/file_download/5451ab4b-6671- 4c6e-af4e-74974247527e

Ministry of Natural Resources and Environment. (2012). Kịch bản biến đổi khí hậu, nốc biển dâng cho Việt Nam (Climate change and sea level rise scenarios for Vietnam). Hanoi, Vietnam: Author.

Moser, S. C., & Ekstrom, J. A. (2010). A framework to diagnose barriers to climate change adaptation. *Proceedings of the National Academy of Sciences, 107*(51), 22026– 22031. doi:10.1073/pnas.1007887107

O'Brien, K. L., Eriksen, S., Schjolden, A., & Nygaard, L. (2004). *What's in a word? Conflicting interpretations of vulnerability in climate change research* (CICERO Work- ing Paper 2004:04). Oslo, Norway: Center for International Climate and Environ- mental Research (CICERO). Retrieved from http://www.cicero.uio.no/media/2682 .pdf

O'Brien, K. L., & Wolf, J. (2010). A values-based approach to vulnerability and adapta- tion to climate change. *Wiley Interdisciplinary Reviews: Climate Change, 1*(2), 232–242. doi:10.1002/wcc.30

Oxfam. (2008). *Evaluation of participatory disaster preparation and mitigation project in Tien Giang and Dong Thap Provinces, Vietnam* (Oxfam GB programme evaluation full report). Oxford, United Kingdom: Author. Retrieved from http://oxfamilibrary .openrepository.com/oxfam/bitstream/10546/119446/1/er-disaster-preparation-vietna m-010608-en.pdf

Schwab, M. (2014). *Value and nature of risk-related strategies—An evaluation of rural coping and adaptation in response to changing water-related risks in the Vietnamese Mekong Delta* (Doctoral dissertation, Universität Bonn, Bonn, Germany). Retrieved from http://hss.ulb.uni-bonn.de/2014/3759/3759.pdf

SenGupta, S., Hopson, R., & Thompson-Robinson, M. (2014). Cultural competence in evaluation: An overview. *New Directions for Evaluation, 102,* 5–19. doi:10.1002/ev.112

Stern, N. H. (2007). *The economics of climate change: The Stern review.* London, United Kingdom: Cambridge University Press.

Turner, B. L., Kasperson, R. E., Matson, P. A., McCarthy, J. J., Corell, R. W., Christensen, L., . . . Schiller, A. (2003). A framework for vulnerability analysis in sustainability sci- ence. *Proceedings of the National Academy of Sciences, 100*(14), 8074–8079. Retrieved from http://www.ncbi.nlm.nih.gov/pmc/articles/PMC166184/

United Nations Framework Convention on Climate Change (UNFCCC) (2010). *Syn- thesis report on efforts undertaken to monitor and evaluate the implementation of adap- tation projects, policies and programmes and the costs and effectiveness of completed projects, policies and programmes, and views on lessons learned, good practices, gaps and*

needs. FCCC/SBSTA/2010/5. Bonn, Germany: Author. Retrieved from http://unfccc
.int/resource/docs/2010/sbsta/eng/05.pdf
Werlen, B. (1993). *Society, action and space: An alternative human geography* (G. Walls,
Trans.). London, United Kingdom: Routledge.
Wisner, B., Blaikie, P., Cannon, T., & Davies, I. (2004). *At risk: Natural hazards, people's
vulnerability and disasters*. London, United Kingdom: Routledge.
World Bank. (2010). *Economics of adaptation to climate change: Vietnam*. Washington,
DC: Author. Retrieved from http://documents.worldbank.org/curated/en/2010/01/
16441103/vietnam-economics-adaptation-climate-change

DUNJA KRAUSE *is associate expert at the United Nations Research Institute for
Social Development (UNRISD), Switzerland.*

MARIA SCHWAB *is research associate at the Helmholtz-Zentrum for Materials
and Coastal Research, Germany.*

JÖRN BIRKMANN *is heading the Institute for Spatial and Regional Planning, University of Stuttgart, Germany.*

McKinnon, M. C., & Hole, D. G. (2015). Exploring program theory to enhance monitoring and evaluation in ecosystem-based adaptation projects. In D. Bours, C. McGinn, & P. Pringle (Eds.), *Monitoring and evaluation of climate change adaptation: A review of the landscape. New Directions for Evaluation, 147*, 49–60.

3

Exploring Program Theory to Enhance Monitoring and Evaluation in Ecosystem-Based Adaptation Projects

Madeleine C. McKinnon, David G. Hole

Abstract

Monitoring and evaluation (M&E) for ecosystem-based adaptation (EbA) projects is in its infancy. Specific M&E challenges lend EbA projects to the use of program theory. In this article, we will examine how two theory-based tools— theory-of-change models and evidence synthesis—can be used to instruct informative EbA M&E through characterizing pathways to impact, identifying causal mechanisms, distinguishing relevant indicators, and recognizing areas of uncertainty, particularly in data-poor environments where intuition and anecdote often substitute for evidence. It will be framed in the context of two EbA case studies, implemented by our organization, in which we will bring together elements from the development of theory-of-change models and provide examples of evidence synthesis underpinning a causal linkage. We argue that such an approach is essential for EbA projects to enable rapid learning and foster a culture of adaptive management. © 2015 Wiley Periodicals, Inc., and the American Evaluation Association.

E cosystem-based adaptation (EbA) is an evolving discipline, nested within the broader field of climate change adaptation. Although it has no strict definition, it is broadly accepted to "include the sustainable management, conservation and restoration of ecosystems to provide services that help people adapt to the adverse impacts of climate change" (Convention on Biological Diversity [CBD], 2009). Given the often-significant overlap between people's dependence on ecosystems and their vulnerability to the impacts of climate change, coupled with the potential for EbA to provide significant biodiversity cobenefits alongside societal adaptation, it is receiving growing attention from both the conservation and development communities (Jones, Hole, & Zavaleta, 2012). Of particular importance are questions about its effectiveness (Doswald et al., 2014), both in terms of its ability to reduce specific vulnerabilities or increase the overall resilience of people to climate change impacts, as well as its cost effectiveness in comparison with other adaptation options. Given that EbA might represent the only easily accessible adaptation option for many poor or rural communities where livelihoods are often particularly tied to natural resources (Jones et al., 2012; Sachs et al., 2009), informative M&E and a strong evidence base are essential to minimize the risk of implementing ineffective interventions that fail to support adaptation.

Monitoring and evaluation of EbA is in its infancy. Although there is a considerable body of literature to draw on from the development field in particular and the climate change adaptation and resilience (CCAR) community specifically, the complexities surrounding the measurement of CCAR project performance are compounded further in EbA. Specific challenges to EbA M&E include (a) complexity of multiobjective strategies that encompass social and biophysical goals (e.g., riverine restoration resulting in reduced asset loss as flood frequency increases under climate change); (b) long-term horizons required to observe social and environmental change (e.g., ecological restoration can take many years); and (c) the level of uncertainty surrounding current tools and approaches (e.g., ecosystem service models) and targeted monitoring of the application of these approaches on the ground. Broader challenges also include the lack of consistent indicators to quantify causal effects; limited availability of relevant data sets; poor characterization of causal pathways linking EbA projects to social and

Funding for this work was provided by the International Climate Initiative (ICI) of the German Federal Ministry for the Environment, Nature Conservation, Building and Nuclear Safety (BMUB), as a component of the project "Ecosystem-based Adaptation in marine, terrestrial and coastal regions as a means of improving livelihoods and conserving biodiversity in the face of climate change." The German Federal Ministry for the Environment, Nature Conservation, Building and Nuclear Safety (BMUB) supports this initiative on the basis of a decision adopted by the German Bundestag. Additional support was provided by the Gordon and Betty Moore Foundation (Grant No. 3519). We are grateful for input and data for case studies from Amanda Bourne and Maria Josella Pangilinan and graphical design support from Gareth Wishart.

biophysical outcomes; and a dearth of scientific evidence underpinning these linkages, as well as critical appraisal of the existing evidence base.

The above challenges lend EbA to program theory-based approaches, such as theory-of-change (ToC) models (White, 2009) that are being increasingly promoted as essential tools for articulating the relationship between adaptation interventions and outcomes (Bours, McGinn, & Pringle, 2014; Conservation International, 2012). In this article, we will examine how two tools—ToC models and evidence synthesis—can be used to instruct informative EbA M&E through characterizing pathways to impact, identifying causal mechanisms, distinguishing relevant indicators, and recognizing areas of uncertainty, particularly in data-poor environments where intuition and anecdote often substitute for evidence.

Information generated by such M&E approaches is valuable for institutional learning that can be applied in adaptive management and future decision making. This article will be framed in the context of two EbA case studies in which we will bring together elements from the development of ToC models and provide examples where evidence synthesis underpins a causal linkage. The case studies are taken from a Conservation International (CI) project that is implementing EbA, with funding from the International Climate Initiative (ICI) of the German Federal Ministry for the Environment, Nature Conservation, Building and Nuclear Safety (BMUB). Our aim is to demonstrate the merits of combining program theory and evidence synthesis to inform M&E strategies for EbA and to highlight existing knowledge gaps.

Importance of Program Theory To Inform EbA Project Archetypes and M&E Design

The potential scope of EbA to help reduce people's vulnerability to a range of climate change impacts is broad, encompassing diverse socioeconomic sectors and hence socio-ecological contexts (Bours et al., 2014; Jones et al., 2012; United Nations Framework Convention on Climate Change [UN-FCCC], 2011). However, the still-nascent state of EbA research and application means practical examples and lessons learned from M&E of EbA projects is lacking. Doswald et al. (2014), for example, found that only 10% of EbA-relevant articles in the peer-review or grey literature had any relevance to project-related monitoring and what existed was high level and incomplete. As a result, EbA generally lacks systematic archetypes or classes of interventions, (e.g., standard activities associated with capacity building or incentive-based programs (Funnell & Rogers, 2011), which can help project teams identify basic elements of the program theory, represent it in a conceptual model or ToC, and outline the M&E needs to measure the outcomes and impacts of their model.

The pathways within a ToC model describe how hypothesized cause-and-effect relationships between elements of the model will bring about desired results (Funnell & Rogers, 2011). Hence, a key benefit of a ToC approach is that it creates the enabling conditions for adaptive management during project implementation. It obliges a project team, ideally in partnership with relevant stakeholders (e.g., communities or local government), to map out the "logic" underpinning a causal pathway connecting project activities to often-complex networks of outputs and outcomes and finally to multifaceted socio-ecological goals and, crucially, the mechanisms that underpin each step in the causal pathway. The latter is particularly important in EbA projects, in which causality is often presumed rather than underpinned by evidence. Data are often lacking, for example, to explicitly link ecosystem service provision under climate change to reductions in the vulnerability of people or communities (Doswald et al., 2014 and see next section).

This step-by-step mapping then allows the project team to identify optimal indicators at key points in the causal pathway both for observing periodic changes (monitoring) and judging effectiveness (evaluation), and begins to address a major issue common to many CCAR projects—namely, time frames that extend well beyond a project's management cycle (Bours et al., 2014). Many EbA projects utilize interventions that seek to restore or rehabilitate one or more ecosystem components. Hence, many of the intended ecological and social outcomes will not manifest themselves for many years, even a decade or more, after the project ends. Mapping a ToC allows a project team to identify both short-term indicators (focused on key outputs or short-term outcomes) for reporting on progress during the project's lifespan and longer-term indicators that track the core ecological and social outcomes underpinning the project goal. It can also highlight differences in the temporal distribution of costs and benefits, a critical factor underpinning project cost effectiveness. Further, assumptions at each step can be identified and articulated. These may result from uncertainties surrounding climate change impacts, the precise mechanisms governing a biophysical or behavioral change, or they may reflect exogenous factors outside the project's control that could nevertheless reduce or even overwhelm the positive effect of the intervention.

Thus, the systematic mapping of causal pathways, mechanisms and attendant assumptions addresses a critical need within the EbA field—namely to enhance project-level monitoring, evaluation, and subsequent learning (Doswald et al., 2014; Munroe et al., 2012; Spearman & McGray, 2011). ToC models provide future evaluators (operating beyond the project cycle) with a means to appraise the long-term efficacy of an intervention against defined quantitative ecological and social goals. Furthermore, by representing the collective logic of the project design and implementation team, a ToC may be the only means by which complex knowledge and learning are succinctly captured once a project team inevitably breaks up or individuals leave an organization. Yet any EbA ToC is only as robust as the evidence base that underpins it. Overwhelming reliance on personal belief or anecdote significantly reduces the likelihood that a project will achieve its goals (Pullin & Stewart, 2006).

Evidence Synthesis and Review as a Means of Assessing EbA Project Effectiveness

Program theory plays a key role in evidence-based policy and practice by enabling translation of a body of knowledge from specific cases to broader contexts (Funnell & Rogers, 2011). Evidence synthesis is a process of compiling, organizing, and assessing results, or knowledge, from different studies to identify and interpret patterns among results, often involving critical appraisal of the quality and strength of evidence using statistical methods associated with meta-analysis. In the conservation sector a community of practice, called The Collaboration for Environmental Evidence, has emerged in the past 10 years, and it promotes systematic review and synthesis of evidence on the effectiveness of conservation interventions (Pullin & Stewart, 2006). As interest in, and funding for, EbA has increased over the past 5 years, calls to document the impacts of EbA projects robustly have increased (Spearman & McGray, 2011; The Nature Conservancy [TNC], 2009). Such evidence is necessary to demonstrate the social, economic, and biophysical benefits of EbA investments, identify effective interventions, provide information to guide future EbA planning and monitoring, highlight potential trade-offs, and avoid maladaptation (actions that may lead to adverse climate-related outcomes, now or in the future). An example of the latter resulting from inadequate evidence synthesis (together with misaligned policy) is the planting of exotic *Casuarina equisetifolia* in various coastal states in India as bioshields, purportedly to protect against extreme events (tsunamis and storm surge). In many areas these plantations have destroyed native coastal dune systems and their attendant vegetation, which, evidence suggests, provide a more stable and effective form of coastal protection (Bhalla, 2007; Feagin et al., 2010).

Initial efforts to document impacts of EbA have included narratives, specific case studies, and nonsystematic reviews, largely confined to the grey literature (Spearman & McGray, 2011; Tompkins & Adger, 2004). Doswald et al. (2014) published the first systematic review of the evidence base that aimed to empirically synthesize the effectiveness of existing EbA interventions. From over 7,000 articles and reports in peer-reviewed publications and grey literature, the review documented outcomes from just 164 studies, reflecting in part the paucity of studies that have systematically monitored and evaluated outcomes. Although measures showed generally positive results of EbA projects in meeting goals, their review suggests two key trends: (a) substantial gaps in the evidence base, and (b) an emphasis on measuring short-term outputs. The first issue requires further investment in M&E, but also the ability of projects to connect evidence from different disciplines (e.g., disaster risk reduction or ecosystem service assessment), in a coherent way and where possible, to identify relevant existing evidence to inform EbA projects. The second requires longer time frames to observe and measure EbA outcomes, but also the ability of projects to predict effects of intermediate results. As evidence is compiled, the next step will be translating these trends and observations into practical guidance to inform what interventions should be applied in what contexts and to effect which populations or groups.

Application of Program Theory-Based Tools in Two Case Studies

In 2010, CI received funding from ICI to implement a 5-year, $6 million project seeking to characterize vulnerability to climate change within three diverse socioecological systems in the Philippines, South Africa, and Brazil and to identify appropriate EbA interventions that would demonstrably reduce the vulnerability of local communities. Interventions ranged from mangrove restoration to reduce the vulnerability of coastal communities in the Philippines to tropical cyclones, to local capacity building and technology transfer in South Africa to restore rangeland health in support of livelihood security under climate change.

For each intervention, ToC models were developed by the project team through a participatory workshop process, to articulate the underlying program theory and map out the complex linkages and relationships between case-study goals and required activities (i.e., the causal pathways). Supporting data were derived from an initial comprehensive vulnerability assessment process for each case study. Each model articulated multiple pathways composed of short-term outputs and outcomes predicted to be observed during the project time frame, which represent requisite steps toward achieving longer-term outcomes and ultimately an overarching (ideally) quantitative, socioecological goal. Assumptions were documented alongside each step in the pathway (not shown). Furthermore, sources of evidence, including systematic reviews, case studies, and reports, were identified for key linkages within the model. The ToC approach was critical in helping the project team identify both short-term indicators for reporting on progress during the project's lifespan, and indicators for longer-term ecological and social outcomes, manifesting postproject, that will be monitored as a result of CI's (and partners) long-term presence at the relevant sites, and used to evaluate long-term impacts.

An example of a causal pathway from the Philippines case study is shown in Figure 3.1 to illustrate the step-by-step logic for a mangrove restoration and rehabilitation intervention in which ecosystem management activities contribute to the achievement of reduced vulnerability of coastal communities facing more frequent or powerful storm surges under climate change. The pathway maps the project team's step-by-step logic toward achieving the project goal based upon a suite of interventions, e.g.,

Figure 3.1. Causal Pathway Within a Theory-of-Change Model for a Mangrove Rehabilitation and Restoration Intervention in the Philippines

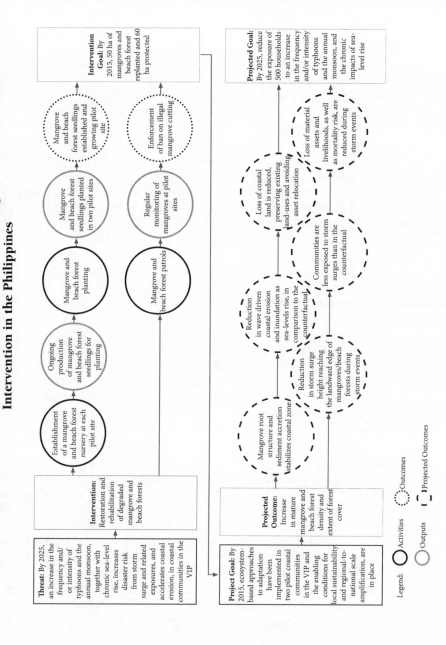

Threat: By 2025, an increase in the frequency and/or intensity of typhoons and the annual monsoon, together with chronic sea-level rise, increases disaster risk from storm surge and related exposures, and accelerates coastal erosion, in coastal communities in the VIP

Intervention: Restoration and rehabilitation of degraded mangrove and beach forests

Establishment of a mangrove and beach forest nursery at each pilot site

Ongoing production of mangrove and beach forest seedlings for planting

Mangrove and beach forest planting

Mangrove and beach forest seedlings planted in two pilot sites

Mangrove and beach forest seedlings established and growing pilot site

Intervention Goal: By 2015, 50 ha of mangroves and beach forest replanted and 60 ha protected

Mangrove and beach forest patrols

Regular monitoring of mangroves at pilot sites

Enforcement of ban on illegal mangrove cutting

Project Goal: By 2015, ecosystem-based approaches to adaptation have been implemented in two pilot coastal communities in the VIP and the enabling conditions for local sustainability and regional-to-national scale amplification, are in place

Projected Outcome: Increase in mature mangrove and beach forest density and extent of forest cover

Mangrove root structure and sediment accretion stabilizes coastal zone

Reduction in storm surge height reaching the landward edge of mangroves/beach forests during storm events

Reduction in wave driven coastal erosion and inundation as sea-levels rise, in comparison to the counterfactual

Loss of coastal land is reduced, preserving existing land-uses and avoiding asset relocation

Communities are less exposed to storm surges than in the counterfactual

Loss of material assets and livelihoods, as well as mortality risk, are reduced during storm events

Projected Goal: By 2025, reduce the exposure of 500 households to an increase in the frequency and/or intensity of typhoons and the annual monsoon, and the chronic impacts of sea-level rise

Legend: ◯ Activities ◯ Outputs ◌ Outcomes ◌ Projected Outcomes

actions designed to bring about changes to biophysical variables, capacity building, and the enabling environment. Figure 3.1 describes only the biophysical intervention. The projected goal reflects long-term outcomes of all interventions and is likely not being achieved during the project lifespan. Examples of hypothesized cause-and-effect relationships reflected in the Philippines causal chain include "if a ban on illegal mangrove cutting is enforced, then mangrove seedlings will be established and conserved," or "if wave-driven coastal erosion is reduced, then coastal land is maintained." These linkages reflect program theories about ecological function, conservation management, or environmental regulatory processes that should be supported by an evidence base. A key hypothesized relationship is that mangrove ecosystems can reduce the height of storm surges that would otherwise inundate coastal areas. This hypothesis is strongly supported by syntheses of empirical evidence (e.g., McIvor, Möller, Spencer, & Spalding, 2012). However, the mechanism by which mangroves reduce wave height is a function of multiple factors including species type, stand and root structure, bathymetry, and wave condition (McIvor et al., 2012). Hence, not all mangroves necessarily produce the same wave-dampening effect. A robust understanding of the local biophysical conditions and socioecological context is therefore critical. A major challenge for EbA projects more broadly is that although evidence exists to support the hypothesis linking mangroves and wave height, many other hypotheses are not yet tested with empirical evidence.

Multidimensional objectives of EbA projects also involve understanding social theories related to human behavior, learning, and diffusion of new ideas. In CI's case study in South Africa, for example, the model highlights a hypothesis that training workshops on climate-smart agricultural practices will improve the capacity of local farmers to implement such techniques (Figure 3.2). However, the causal mechanism by which training affects farmers' knowledge is a function of how people learn and use new information. Although the evidence base in education and capacity building is substantial, it generally lacks specificity for particular types of capacity and various socioeconomic contexts. Many context-specific variables will affect the occurrence of desired effects from social norms, trust between farmers and trainers, relevant incentives, and necessary reenforcement of new knowledge. This causal linkage is therefore more reliant on preconceived beliefs rather than evidence about whether capacity building will be achieved as a result of the planned training activity. M&E is therefore critical to track progress and adjust intervention design as needed.

In these two examples, we have briefly explored causal pathways, program theory, assumptions, and empirical evidence behind different models. Each ToC model underpinning the case studies is replete with multiple pathways, each navigating a fine line between sources of evidence and areas of uncertainty. M&E is therefore critical. We need to understand how, why, and where EbA works. Our approach—the use of theory-based ToC

Figure 3.2. Section of a Causal Pathway for a Capacity-Building Intervention in South Africa (Full Causal Pathway Not Shown)

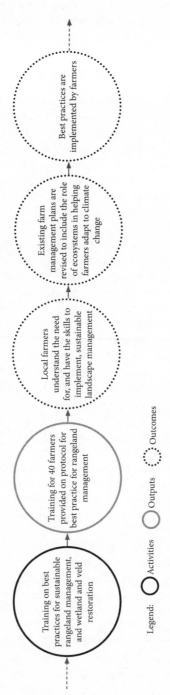

Training on best practices for sustainable rangeland management, and wetland and veld restoration

Training for 40 farmers provided on protocol for best practice for rangeland management

Local farmers understand the need for, and have the skills to implement, sustainable landscape management

Existing farm management plans are revised to include the role of ecosystems in helping farmers adapt to climate change

Best practices are implemented by farmers

Legend: ◯ Activities   ◯ Outputs   ◌ Outcomes

Note: See Figure 3.1 for a description of the various elements.

models of projected changes complemented by existing evidence—is an optimal means to achieve these ends for both current project teams, as well as future managers.

Discussion

In this article, we have described two tools—ToC models and evidence synthesis—that inform planning and M&E of EbA projects. When used together, they have the potential to expedite learning on which interventions are most effective, what indicators to measure, and how to deal with uncertainty in decision making. This approach recognizes specific challenges in EbA projects including long-term horizons, deep uncertainty, and the complexity of multiobjective projects. Significant risks exist for EbA projects that do not take a systematic and evidence-based approach to their design and delivery. At a minimum, projects will be inefficient or difficult to replicate. More seriously, they will risk wasting valuable resources investing in activities that are ineffective or produce unintended, possibly adverse, outcomes. More broadly, a ToC approach will also help EbA practitioners to embrace nascent conceptual advances that may further strengthen project design, implementation and adaptive management, such as 'adaptation pathways,' that seek to identify key future decision points where projected outcomes need to be reappraised in light of realized climate change, biophysical tipping points and thresholds, as well as shifts in societal values, rules and knowledge (Werners et al., 2014; Wise et al., 2014).

The importance of ToC models is emphasized throughout articles of this issue. Building upon recent guidance and reviews in the CCAR field, ToC models are explicitly highlighted in chapters authored by Adler, Wilson, Abbot, and Blackshaw (this issue), and Faulkner, Ayers, and Huq (this issue). Within our organization (CI), for example, there is a growing interest to use ToC models to communicate our approach and to understand how our strategies contribute to global conservation and development goals. ToCs enable us to embed our actions in systematic, long-term logic, scale up our monitoring efforts, and map out long-term projected outcomes when direct attribution is unclear. Challenges to monitoring at different scales were raised by other colleagues from CI in a previous issue of *New Directions for Evaluation* focused on environmental evaluation (Kennedy, Balasubramanian, & Crosse, 2009). A consistent element between this current article and that of Kennedy et al. is the need for scalable monitoring frameworks, where site and project-level outcomes can be aggregated for the purpose of programmatic and institutional reporting. Less explicit among other articles in this issue and the frameworks described therein, is the consideration or integration of existing evidence into models, or how pathways were characterized, for example, quality or strength of linkages relative to available empirical data. The CI case studies are some of the first attempts to use such an approach purposefully for an EbA project. As an implementing organization with decades-long field presence, this approach is advantageous for new project teams and partners to understand the collective logic and assumptions surrounding interventions selected by past projects, to update with new data and values, and check progress against long-term goals.

Mapping causal pathways toward adaptation goals is a positive step toward promoting the culture of adaptive management necessary for EbA projects to rapidly learn, improve, and ultimately be more effective. Three mechanisms can contribute to achievement of this goal based upon actions by institutions, donors, and policy makers. The first involves the use of consistent monitoring systems by implementing organizations, such as CI, which are grounded in program theory. This begins with building ToC models and identifying relevant standard indicators, which are then used to gather baseline data necessary for M&E. The second involves financial support and other incentives from donors for longer-term monitoring, or otherwise, commitment by donors to measure results of their investments independently. Third, as national governments commit to meaningful climate policy targets and adopt national adaptation plans to achieve these, policy frameworks should integrate program theory to ensure robust evidence-based policy and decision making, along with greater support for standardized national monitoring systems.

References

Bhalla, R. S. (2007). Do bio-shields affect tsunami inundation? *Current Science, 93,* 831–833.

Bours, D., McGinn, C., & Pringle, P. (2014). *Guidance note 3: Theory of change approach to climate adaptation planning.* Phnom Penh, Cambodia: SEA Change CoP, and Oxford, United Kingdom: UKCIP. Retrieved from http://www.ukcip.org.uk/wordpress/wp-content/PDFs/MandE-Guidance-Note3.pdf

Conservation International. (2012). *Constructing theories of change for ecosystem-based adaptation projects: A guidance document.* Arlington, VA: Author. Retrieved from http://sp10.conservation.org/Documents/CI_IKI-ToC-Guidance-Document.pdf

Convention on Biological Diversity (CBD). (2009). *Connecting biodiversity and climate change mitigation and adaptation: Report of the second ad hoc technical expert group on biodiversity and climate change* (CBD technical series no. 41). Montreal, Canada: CBD Secretariat.

Doswald, N., Munroe, R., Roe, D., Giuliani, A., Castelli, I., Stephens, J., & Reid, H. (2014). Effectiveness of ecosystem-based approaches for adaptation: Review of the evidence base. *Climate and Development.* doi:10.1080/17565529.2013.867247

Faulkner, L., Ayers, J., & Huq, S. (2015). Meaningful measurement for community-based adaptation. *New Directions for Evaluation, 147,* this issue.

Feagin, R. A., Mukherjee, N., Shanker, K., Baird, A. H., Cinner, J., Kerr, A. M., & Dahdouh-Guebas, F. (2010). Shelter from the storm? Use and misuse of coastal vegetation bioshields for managing natural disasters. *Conservation Letters, 3*(1), 1–11.

Funnell, S. C., & Rogers, P. J. (2011). *Purposeful program theory: Effective use of theories of change and logic models.* Hoboken, NJ: John Wiley & Sons.

Jones, H. P., Hole, D. G., & Zavaleta, E. S. (2012). Harnessing nature to help people adapt to climate change. *Nature Climate Change, 2,* 504–509.

Kennedy, E. T., Balasubramanian, H., & Crosse, W. E. M. (2009). Issues of scale and monitoring status and trends in biodiversity. *New Directions for Evaluation, 122*, 41–51.

McIvor, A. L., Möller, I., Spencer, T., & Spalding, M. (2012). *Reduction of wind and swell waves by mangroves.* Cambridge, United Kingdom: The Nature Conservancy, and Wageningen, The Netherlands: Wetlands International. Retrieved from http://www.wetlands.org/Portals/0/publications/Report/reduction-of-wind-and-swell-waves-by-mangroves.pdf

Munroe, R., Roe, D., Doswald, N., Spencer, T., Moller, I., Vira, B., & Stephens, J. (2012). Can ecosystem-based approaches help people adapt to climate change? *Environmental Evidence, 1*(13), 1–11.

Pullin, A. S., & Stewart, G. B. (2006). Guidelines for systematic review in conservation and environmental management. *Conservation Biology, 20*(6), 1647–1656.

Sachs, J. D., Baillie, J.E.M., Sutherland, W. J., Armsworth, P. R., Ash, N., Beddington, J.,... Jones, K. E. (2009). Biodiversity conservation and the Millennium Development Goals (MDGs). *Science, 325*, 1502–1503.

Spearman, M., & McGray, H. (2011). *Making adaptation count: Concepts and options for monitoring and evaluation of climate change adaptation* [Manual]. Eschborn, Germany: Deutsche Gesellschaft für Internationale Zusammenarbeit (GIZ) GmbH, and Bonn, Germany: Bundesministerium für wirtschaftliche Zusammenarbeit und Entwicklung (BMZ), and Washington, DC: World Resources Institute (WRI). Retrieved from http://www.wri.org/publication/making-adaptation-count

The Nature Conservancy (TNC). (2009). *Adapting to climate change: Ecosystem-based approaches for people and nature.* Arlington, VA: Author.

Tompkins, E. L., & Adger, W. N. (2004). Does adaptive management of natural resources enhance resilience to climate change? *Ecology & Society, 9*, 1–14.

United Nations Framework Convention on Climate Change. (2011). *Ecosystem-based approaches to adaptation: Compilation of information.* Durban, South Africa: Subsidiary Body for Scientific and Technological Advice.

Werners, S., Pfenninger, S., Slobbe, E. v., Haasnoot, M., Kwakkel, J., & Swart, R. J. (2014). Thresholds, tipping and turning points for sustainability under climate change. *Current Opinion in Environmental Sustainability, 5*, 334–340.

White, H. (2009). *Theory-based impact evaluation: Principles and practice* (Working Paper No. 3). New Delhi, India: International Initiative for Impact Evaluation (3ie). Retrieved from http://www.3ieimpact.org/media/filer_public/2012/05/07/Working_Paper_3.pdf

Wise, R. M., Fazey, I., Stafford Smith, M., Park, S. E., Eakin, H. C., Archer Van Garderen, E. R. M., & Campbell, B. (2014). Reconceptualising adaptation to climate change as part of pathways of change and response. *Global Environmental Change, 28*, 325–336.

MADELEINE C. MCKINNON *is a senior director for Monitoring & Evaluation at Conservation International (CI), USA, focusing on measuring and reporting CI's institutional achievements across its global programs.*

DAVID HOLE *is a senior director for Global Synthesis at Conservation International (CI), USA, focusing on the contribution of natural capital to human wellbeing.*

Adler, R., Wilson, K., Abbot, P., & Blackshaw, U. (2015). Approach to monitoring and evaluation of institutional capacity for adaptation to climate change: The case of the United Kingdom's investment to Ethiopia's climate-resilient green economy. In D. Bours, C. McGinn,
& P. Pringle (Eds.), Monitoring and evaluation of climate change adaptation: A review of the
landscape. New Directions for Evaluation, 147, 61–74.

4

An Approach to Monitoring and Evaluation of Institutional Capacity for Adaptation to Climate Change: The Case of the United Kingdom's Investment in Ethiopia's Climate-Resilient Green Economy

Rebecca Adler, Kirsty Wilson, Patrick Abbot, Ursula Blackshaw

Abstract

The design and implementation of evaluation methods assessing the impact of
the United Kingdom's climate finance on the Ethiopian Government's institutional capacity for implementation of the national climate change strategy is described. The article focuses on capacity-assessment tools that support evidence-
based self-assessments of progress toward politically determined targets. We
highlight the strengths of the capacity-assessment tools in building ownership of
monitoring data among implementers, providing rapid feedback to inform management of development programs, and the potential to create windows of opportunity to tackle long-standing institutional blockages. We also note challenges
in relation to the skills required to implement the evaluation methods effectively,
the limitations in attributing change to a particular intervention, and potential
biases. Finally, we explain the scope for use of the tools in other contexts. ©
2015 Wiley Periodicals, Inc., and the American Evaluation Association.

This article discusses the design and implementation of a monitoring and evaluation (M&E) methodology to assess the impact of the United Kingdom's climate finance in Ethiopia on policy and institutional structures at national and subnational levels. Although the evaluation methodology applied is discussed, the focus of this article is largely on one component—a set of tools that were developed to measure changes in institutional structures and capacities. In particular, these tools enabled the evaluators to understand whether changes in institutional structures and capacities were being made and whether these were contributing to the Government of Ethiopia's long-term goal of a transformative shift to a climate-resilient, low-carbon economy. The tools were tailored for the Ethiopian context but drew on other approaches that measure general organizational capacity, as well as those for climate-specific capacity assessment. Ease of application and ability to collect information on key criteria were key design considerations. The need to generate both qualitative and quantitative data through participatory processes was also considered.

Although it is a least-developed country, Ethiopia was also the world's 12th-fastest-growing economy in 2012 (World Bank, 2013). Key to Ethiopia's national development plan is the target of becoming a middle-income country by 2025 and an ambitious climate goal—the Climate-Resilient Green Economy (CRGE)—of achieving economic targets with no net growth in carbon emissions (Federal Democratic Republic of Ethiopia [FDRE], 2011). Ethiopia is one of the few developing countries to have embedded climate-related objectives into their mainstream development plan (Jones & Carabine, 2013).

Climate change has the potential to have negative impacts in Ethiopia. For example, agriculture, a critical part of the Ethiopian economy and the primary livelihood for 85% of the population (FDRE, 2002) is highly sensitive to climate variability. Although national models and missing data obscure subnational differences in variability and trends, there is evidence of increasing mean annual temperatures and declining seasonal rains (Conway & Schipper, 2011).

Climate change also poses opportunities to revisit past efforts to catalyze sustainable development and reduce both human and infrastructural vulnerabilities. This requires consideration of how mitigation, adaptation, and development policies can best be balanced (Klein, Schipper, & Dessai, 2005). Ethiopia's response is to embed climate-resilience actions within a wider set of policy responses aimed at economic transformation (FDRE, 2011). However, there are a number of uncertainties associated with this approach. These include trade-offs between competing social, economic, and environmental goals, availability of international finance, and questions about the most appropriate and effective policies and institutional arrangements for achieving the desired goals (Collier & Venables, 2012).

Given these uncertainties and the novelty of this approach, stakeholders in Ethiopia have adopted a "learning by doing" approach to climate policy (Bass, Ferede, Fikreyesus, & Wang, 2013), which is consistent with other adaptive management approaches to complex policy processes (Andrews, Pritchett, & Woolcock, 2012; Swanson et al., 2009). Not only is the pathway for adaptation to climate change untested, but the literature indicates that adapting to climate change must be continuous, requiring a combination of incremental and transformative actions that lead to new institutional and policy arrangements (Kates, Travis, & Wilbanks, 2012; Levine, Ludi, & Jones, 2011). Institutions are shown to be important in influencing adaptation to climate change, in particular by structuring climate change impacts, shaping adaptation objectives, and channeling resources (Agrawal, 2010; Dixit, McGray, Gonzales, & Desmond, 2012).

The United Kingdom is supporting this "learning by doing" approach through two large-scale programs funded by the International Climate Fund (ICF), an interministerial fund, and with ongoing dialogue between U.K. Department For International Development (DFID) advisors, the program managers, and multiple actors within the Ethiopian Government. Both DFID and the Government of Ethiopia are interested in catalyzing the transformative shift to a climate-resilient, low-carbon economy. Changes in policy and financial instruments, institutional arrangements, structures, and capacities, and projects aimed at implementing the CRGE policy objectives are needed.

DFID commissioned LTS International to deliver on-going monitoring and evaluation and capture real-time learning from these investments. The first program—the Strategic Climate Institutions Programme—created a capacity-building challenge fund and a private-sector finance mechanism. The second—the Climate High Level Investment Programme—provided funds aimed at catalyzing sectoral and strategic mainstreaming in six CRGE priority sectors.

Approach and Tools

A theory-based M&E methodology was designed to test the assumptions underlying the theory of change for these two programs and to capture aspects of organizational and institutional capacity that were identified as important by DFID and the Ethiopian Government. Necessary improvements were jointly identified in three complementary capacities within the Government of Ethiopia systems. First, sector line ministries, (and relevant private sector and civil society organizations) were required to develop the capacity to plan and implement climate-resilient green development actions. Second, government line ministries needed to be able to apply lessons learned from these pilots to mainstream development plans and investments. And third, the government's national climate finance mechanism (the CRGE Facility) would need to make effective investment decisions and use evidence for leveraging funds.

Three tools were developed to monitor changes in these three elements of the theory of change—they are the Organizational Capacity-Assessment Matrix (OCAM), Mainstreaming Capacity-Assessment Matrix (MCAM), and Decision-Making Capacity-Assessment Matrix (DCAM). (The complete matrices, including questions asked of respondents and qualitative coding schemes, are available from the authors.) The tools, which are the focus of this article, are complemented by other data-collection methods and analysis of progress. The tools did not solely focus on elements related to the ICF support. In the case of the OCAM, the tool identified and measured the full set of required competencies for climate policy implementation. For example, if the ICF investments strengthened capacity in one area, but complementary government efforts were not applied in new procedures for district level staff, the tool would identify this as a missing link in the theory of change. The tool had 24 subcomponents under its four capacity areas. For both the MCAM and DCAM, to reduce the time required for the assessments, the focus was more narrowly defined on issues of interest to the DFID program. A summary of the main features of the three tools is found in Table 4.1.

These tools were developed following a review of relevant literature and tools, including the National Capacity Adaptive Framework (World Resources Institute [WRI], 2009), the Tracking Adaptation and Measuring Development approach (Brooks et al., 2013), the Organizational Capacity-Assessment Tool (McKinsey, 2014) and ICF guidance (e.g., ICF, 2011, 2013).

The tools are based on a consistent matrix format with interview questions resulting in ordinal scoring and qualitative data. Questions are asked in interviews with up to three local stakeholders from the same government department. Applicable to any governance level, from national to grassroots, the tools are primarily applied at the national and regional levels. Respondents are asked to assess the capacity of their organization to address the relevant subcomponent and to provide evidence to support this assessment.

To illustrate the nature of evidence collected, Table 4.2 illustrates evidence types that were used in the Mainstreaming Capacity Assessment.

A simplified scoring system is applied in the DCAM and MCAM for respondents who scored 0 (none), 1 (partial), or 2 (yes) on each subcomponent based on the availability of evidence required. The OCAM used a scoring scale of 1–4. To facilitate evidence collection, each subcomponent is described, and a question provided to understand capacity within each subcomponent better. Finally, for each subcomponent score, a statement is provided that indicates the evidence and responses that would be required to achieve that score. Table 4.3 illustrates an example of the format of the tool and level of guidance provided.

As the example illustrates, the tools apply a participatory self-assessment approach. Guidance is provided for the assessor on the scoring criteria and the type of evidence required to achieve specific scores. The

Table 4.1. Main Features of the Capacity-Assessment Tools

Tool	Evaluation objects	Components of capacity assessed	Selected indicators measured
Organizational capacity-assessment matrix (OCAM)	Government, CSO and private sector (local and federal)	Planning, monitoring and evaluation (eight subcomponents), program delivery (seven subcomponents), learning (five subcomponents) and communications and influence (four subcomponents).	Organizational capacity to lead CRGE preparation and implementation; level of awareness and understanding of CRGE concepts and planning in regional government; capacity of line ministry and regional bureaus to plan and learn in relation to climate change.
Mainstreaming capacity-assessment matrix (MCAM)	Sector line ministries in six priority sectors; regional climate change focal and economic and finance bureaus in two sample regions.	Planning for mainstreaming (four subcomponents), staff awareness and skills (two subcomponents), safeguards and equity (two subcomponents).	Level of integration of climate change in national and regional planning
Decision-making capacity-assessment matrix (DCAM)	National climate financing facility.	Fiduciary risk management (three subcomponents), technical quality assurance (two subcomponents), innovation and learning (two subcomponents), safeguards and equity (three subcomponents).	CRGE facility decision-making capacity; level of inclusion of safeguards assessment and action in CRGE financed projects and plans

Table 4.2. Evidence Types Used in the Mainstreaming Capacity-Assessment Tool

Capacity component	Evidence types
Planning systems for mainstreaming	Existence of sector/regional CRGE strategy document
	Existence of authoritative body, budgeted and staffed and mandated with coordinating climate change planning and actions
	Routine screening of ministry action plans for climate risks
	Inclusion of CRGE measures in mainstream programs and routine activities
Staff awareness and skills	Availability of staff with adequate knowledge of climate mainstreaming present in relevant directorates of the ministry
	Planned training and development activities to improve staff awareness and skills in relation to CRGE
Safeguards and equity	Awareness of potential trade-offs between resilience, greenhouse gas mitigation, and development and criteria for prioritization
	Documented evidence of the existence and application of a safeguard system that integrates social and environmental risks
	Documented evidence of efforts to improve the contribution of CRGE activities to the promotion of gender equality

assessor and interviewees jointly identify the appropriate score based on the evidence provided in the interview and supplemental written evidence provided or referenced. The findings and evidence are summarized and agreed upon formally with each interviewee at the conclusion of each interview. There is no weighting, as each element of capacity is seen as equally important. Some scores are linked to standalone indicators of particular interest to the DFID programs.

The information was then used in two ways. For the OCAM, the mode of the ordinal scores provides an indication of capacity and was disaggregated by dimension. For the DCAM and MCAM, a cumulative score (an approach used by the ICF mainstreaming indicator scorecard) was used to enable comparison between organizations and *within* the same organizations over time.

The score (or change in score from a previous assessment) was used within the program logframes to measure changes of capacity of these organizations as proxy indicators of progress; changes cannot be solely attributed to the DFID programs. These quantitative scores feed into the overall program logframes and are supplemented by additional indicators that capture additional elements of the process that are more directly attributable to the programs. For example, the MCAM broadly captures the extent

Table 4.3. Example of Tool Scoring Guide (Subcomponent 1—Managing Knowledge and Learning of the Learning Dimension of the OCAM)

Capacity component		Capacity-assessment question	Assessment criteria				Reasons/supporting evidence (~3 bullets)
Component	Sub-component	To what extent . . .	1	2	3	4	
Learning	3.1 Managing knowledge and learning	(a) Is the organization's CRGE knowledge and learning from implementing programs and projects captured?	CRGE knowledge and learning is not captured.	Knowledge and learning is captured for less than 50% of CRGE programs and projects.	Knowledge and learning is captured for 50–75% of CRGE programs and projects.	Knowledge and learning is captured for more than 75% of CRGE programs and projects.	
		(b) Does the organization have a CRGE knowledge management system that systematically records, catalogues, and shares learning with internal and external stakeholders?	The organization does not have a CRGE knowledge management system.	The CRGE knowledge management system is partly designed, but requires further development refinement.	The CRGE knowledge management system has been designed but is not fully operational.	A well-documented knowledge management system ensures that knowledge is systematically recorded, catalogued and shared with internal and external stakeholders.	

Note. This section specifically refers to lessons learned from experience of implementing CRGE, and which can then be shared with others to enable them to apply the lessons.

to which the Ministry of Agriculture developed plans about the climate-resilient green economy, but fails to capture the extent to which the lessons from the DFID program were reflected in a particular program plan. A supplementary logframe indicator tracks this progress. Similarly, although the OCAM captures a general sense of progress in relation to learning (see the example in Table 4.3), the overall evaluation methodology also collects specific qualitative information on the extent to which lessons learned from the United Kingdom's investments are used in other government or civil society programs.

The nature of these assessment processes allows the assessor to probe the credibility of evidence provided by respondents. For example, if a respondent comments that ministerial work plans are routinely screened for climate risk, the interviewer might ask to see documentation of risk assessments or actions taken in response. If a respondent comments that gender advisors exist within the department and are tasked with reviewing CRGE plans, the interviewer can ask if these advisors have influenced any CRGE plans. These shared discussions of evidence also provide a clearer basis for joint agreement of scores.

The assessment tools are used at the beginning of the program to set a baseline score for each organization, and again periodically, as part of annual, mid-term, or final reviews, to allow for an update of progress using the indicators and collect data for specific evaluation questions. Contribution analysis is an integral part of these reviews to analyze the extent to which the United Kingdom's investments played a role in creating the changes measured.

It is worth noting that the selection of capacity elements is political as much as technical, and the design and scope of these tools reflect the outcome of policy engagement between DFID and the Government of Ethiopia. Government stakeholders not directly engaged in these negotiations but who participated in the assessments welcomed an explanation of this process. When the tools were presented to respondents, enumerators explained why particular capacities had been chosen and how the data would be used. Informal feedback from interviewees indicated that they appreciated the structured process and its help in understanding CRGE objectives and their own role within it.

Strengths of the Tools

While tailored specifically to the aspects of organizational and institutional capacity identified by DFID and the Ethiopian Government, the tools also serve as entry points to wider questions of institutional reform—for example, in promoting discussion on civil service reform or the need for greater transparency for safeguard systems.

Historically, donor efforts to push developing country governments to undertake institutional reform have frequently failed (Pritchett, Woolcock,

& Andrews, 2010). However, the nature of the CRGE agenda and the high level of Ethiopian political support created a new space for donor and government dialogue on some previously intractable institutional challenges as the tools go beyond donor demands. For example, there is a renewed interest in strengthening social and environmental impact assessment processes and in increasing transparency about the outcomes of these systems. In addition, institutional arrangements and staffing requirements for supporting high-quality planning within ministries have also received renewed attention as the CRGE agenda bridges economic and environmental priorities.

The fact that the capacity-assessment tools rely upon a self-assessment discussion supported by review of evidence provides a credible basis for dialogue that is not only externally driven. For example, the MCAM highlighted the lack of knowledge among Planning Directorates of the Sector's CRGE strategy as well as the weaker planning skills of Ministry CRGE Units, opening discussions of potential institutional reform. Similarly, as a result of the evidence from the MCAM, the effectiveness of environmental impact assessments as a climate risk tool was questioned.

In particular, the value of the tools as an entry point to pre-existing development challenges around institutional reform is exemplified by the Ethiopian Government's commission of an adapted version of the OCAM to assess the availability of particular capacities at district, regional, and federal sector levels evidence for the design of a large-scale capacity-building program related to climate change. The elements of capacity measured by the government's OCAM overlap with those used in the three monitoring and evaluation tools described, but the outcomes will inform program design. The tool covers core organizational capacity elements such as planning, resource mobilization, and financial management. This demonstrates how climate change can act as an entry point to institutional reform processes that can ultimately address broad and challenging issues that touch upon all aspects of governance.

The simplicity of the capacity-assessment tools is one reason for the government's interest in replicating its design. The level of detail included in the instructions, matrix, and scoring sheet ensures relative consistency of scoring between different assessors. The tools have been developed for use in a 1 to 2 hour interview, minimizing demands on stakeholders. Although the interviews focus on the qualitative evidence of capacity, the evidence-based ordinal scoring of data allows for both qualitative and descriptive quantitative analysis (e.g., correspondence analysis). Finally, the self-assessment approach has increased awareness of line ministries to areas or capacities that are weaker or newer to the organization. This not only supports the DFID investment by measuring its impact, but also contributes to raising awareness among Ethiopian stakeholders about the expected impact of the DFID program. It also provides the recipients a measure of progress of their current status that they can use to improve the focus of their institutional development efforts, because the DFID investment is aligned with governmental CRGE institutional efforts. Because DFID investments aim to have a transformational effect—potentially impacting all aspects of the government's planning—this process has supported both the United Kingdom's and Ethiopian objectives.

Application Challenges

The tools do have limitations to their application. For example, effective qualitative data collection and scoring require interviewing and analytical skills that are not always readily available. This is particularly true in contexts where some stakeholders fear open discussion of government capacity gaps. Training local enumerators to assess the answers provided by government stakeholders, probe for satisfactory evidence, and come to agreement on a score required intensive coaching and practice. Implementing these tools is most effective if the interviewer has prior experience in working with government, critical thinking skills, and a level of seniority, which has implications on the availability of suitable enumerators.

Although the simplicity of the tool limits the time burden on respondents and facilitates data collection by local enumerators, it also limits the number of capacity components that are considered. However, the trade-offs between feasibility and rigor in the design have been managed to produce a cost-effective outcome.

Another challenge is that using a score as a proxy for institutional capacity can result in a reductionist understanding of institutional change processes. In particular, this might risk downplaying political factors and limit nuanced understandings of the factors that enable or block institutional change. The scores are justified by qualitative information that is useful for both DFID and Ethiopian Government program decision makers. For example, the MCAM indicated where staff shortages and poor interdepartmental communication limited mainstreaming progress or technical quality assurance of climate investments. The overall evaluation methodology relies on complementary processes that analyze the qualitative information collected.

The tools alone do not generate strategies to address limitations, but rather generate evidence to support accountability and to identify problematic areas. Decision-making was better supported by complementary lesson learning workshops that offered a more flexible forum for creative problem-solving discussions.

Institutional assessments also face the challenge of individual or organizational interests influencing the assessment scores. In Ethiopia, respondents tended to self-score highly to avoid appearing deficient in their ability to deliver a political priority or to push for low scores in order to capture resources. It is impossible to eliminate all subjectivities from qualitative work, and each application of the tool generates a product that reflects the interests of both interviewer and interviewee. However, building aware-ness of the scope for such biases encouraged enumerators to exercise greater vigilance, to probe for evidence, and to reach agreement on a score based primarily on evidence. Document review and interviews with other stake-holders provided further opportunities to triangulate findings and push back when respondents' self-scoring appeared inaccurate. Documenting the circumstances under which interviews were conducted and external quality assurance of all completed interview notes also helped improve the quality of the data generated, reducing potential bias. Because the purpose of these assessments was not to allocate resources, but to provide information on ca-pacity changes over time, the influence of organizational interests was also limited.

A final challenge was the requirement to attribute changes in institu-tional capacity to DFID's investments and policy engagement. Given the complex nature of institutional and policy change, statistical attribution is not possible. However, the tools can provide useful data for an overall con-tribution analysis. With the use of a process-tracing approach, this analysis examines the evidence in support of DFID's theory of change as well as other plausible theories of change and builds a picture of both necessary and suf-ficient conditions for the changes that have occurred. The interviews used for data collection offer an opportunity to explore the reasons for changes in institutional capacity, but alone would not be sufficient to attribute change to any particular intervention.

Conclusion

Although the tools do have weaknesses, in that the capacities measured are not exhaustive and the simple non-weighted aggregation approach does not provide a full picture of the underlying complexity of progress toward these objectives, their strengths and benefits are significant. They provide a structured and simple mechanism for collecting and analyzing evidence of organizational and institutional capacities required to deliver the CRGE. The resulting data are easy for stakeholders at all levels of government (dis-trict to federal) to understand, interpret, and use for tracking progress and identifying gaps and needs. Using these tools at key intervals provides both readily understood quantitative indicators of the direction and extent of progress made and rich qualitative evidence that can inform Ethiopia's fu-ture actions and DFID's adaptive delivery of its support.

The Government of Ethiopia is able and willing to use the results to-ward institutional reform, partly because they can easily internalize them: The design of these tools reflected the outcome of policy engagement be-tween the two partners, and the self-assessment process ensures that results are not external, donor-driven recommendations.

The tools are uniquely tailored to their context, but the broad elements of the approach and the simplicity of the tools lend themselves to adaptation for other contexts, such as for the evaluation of other institutional change processes or capacity assessments (LTS International, 2014). However, when adapting these tools, evaluators should be aware of the importance of making explicit the political process used in selecting success criteria against which the tools generate scores and the need for complementary processes to ensure the effective use of the rich qualitative data that accompanies scoring.

References

Agrawal, A. (2010). Local institutions and adaptation to climate change. In: R. Mearns & A. Norton (Eds.), *Social dimensions of climate change: Equity and vulnerability in a warming world* (pp. 173–198). Washington, DC: World Bank. Retrieved from https://openknowledge.worldbank.org/handle/10986/2689

Andrews, M., Pritchett, L., & Woolcock, M. (2012). *Escaping capability traps through problem-driven iterative adaptation (PDIA)* (CGD Working Paper No. 299). Washington, DC: Centre for Global Development (CGD). Retrieved from http://ssrn.com/abstract=2102794

Bass, S., Ferede, T., Fikreyesus, D., &Wang, S. (2013). *Making growth green and inclusive.* Paris, France: Organisation for Economic Co-operation and Development (OECD) Publishing. Retrieved from http://www.oecd-ilibrary.org/environment/making-growth-green-and-inclusive_5k46dbzhrkhl-en

Brooks, N., Anderson, S., Burton, I., Fisher, S., Rai, N., & Tellam, I. (2013). *An operational framework for tracking adaptation and measuring development* (Climate Change Working Paper No. 5). London, United Kingdom: International Institute for Environmental Development (IIED). Retrieved from http://pubs.iied.org/pdfs/10038IIED.pdf

Collier, P., & Venables, A. J. (2012). Greening Africa? Technologies, endowments and the latecomer effect. *Energy Economics, 34*(S1), S75–S84.

Conway, D., & Schipper, E.L.F. (2011) Adaptation to climate change in Africa: Challenges and opportunities identified from Ethiopia. *Global Environmental Change, 21*(2011), 227–237.

Dixit, A., McGray, H., Gonzales, J., & Desmond, M. (2012). *Ready or not: Assessing institutional aspects of national capacity for climate change adaptation.* Washington, DC: World Resources Institute (WRI). Retrieved from http://www.wri.org/publication/ready-or-not

Federal Democratic Republic of Ethiopia (FDRE). (2002). *Ethiopia: Sustainable development and poverty reduction program.* Addis Ababa, Ethiopia: Author. Retrieved from http://www.imf.org/external/np/prsp/2002/eth/01/073102.pdf

Federal Democratic Republic of Ethiopia. (2011). *Ethiopia's vision for a climate resilient green economy.* Addis Ababa, Ethiopia: Author. Retrieved from http://www.preventionweb.net/english/professional/publications/v.php?id=24317

International Climate Fund (ICF). (2011). *International Climate Fund (ICF) implementation plan 2011/12–2014/15* [Technical paper]. London, United Kingdom: Author. Retrieved from https://www.gov.uk/government/uploads/system/uploads/attachment_data/file/66150/International_Climate_Fund__ICF__Implementation_Plan_technical_paper.pdf

International Climate Fund. (2013). *Summary of International Climate Fund (ICF) key performance indicators.* London, United Kingdom: Author. Retrieved from https://www.gov.uk/government/uploads/system/uploads/attachment_data/file/253682/ICF-KPI-summary.pdf

Jones, L., & Carabine, E. (2013). *Exploring political and socio-economic drivers of transformational climate policy: Early insights from the design of Ethiopia's climate resilient green economy strategy* (Overseas Development Institute [ODI] Working Paper). London, United Kingdom: ODI. Retrieved from http://www.odi.org/sites/odi .org.uk/files/odi-assets/publications-opinion-files/8617.pdf

Kates, R. W., Travis, W. R., & Wilbanks, T. J. (2012). Transformational adaptation when incremental adaptations to climate change are insufficient. *Proceedings of the National Academy of Sciences, 109*(19). Retrieved from http://10.1073/pnas.1115521109

Klein, R., Schipper, L., & Dessai, S., (2005). Integrating mitigation and adaptation into climate and development policy: Three research questions. *Environmental Science & Policy, 8*(6), 579–588.

Levine, S., Ludi, E., & Jones, L. (2011). *Rethinking support for adaptive capacity to climate change: The role of development interventions* [A report for the Africa Climate Change Resilience Alliance (ACCRA)]. London, United Kingdom: ODI. Retrieved from http:// www.odi.org/publications/6213-accra-adaptive-capacity-development-interventions

LTS International. (2014). *Regional capacity assessment tool for the climate resilient green economy capacity development program* (Unpublished Report). Edinburgh, Scotland: Author.

McKinsey and Company. (2014). *Organizational Capacity Assessment Tool V2.0* [Web page]. Retrieved from http://mckinseyonsociety.com/ocat/

Pritchett, L., Woolcock, M., & Andrews, M. (2010). Capability Traps? The Mechanisms of Persistent Implementation Failure - Working Paper 234. Washington, DC: Center for Global Development (CGD). Retrieved from http://www.cgdev.org/publication/ capability-traps-mechanisms-persistent-implementation-failure-working-paper-234

Swanson, D., Barg, S., Tyler, S., Venema, H., Tomar, S., Bhadwal, S., Nair, S., Roy, D., & Drexhage, J. (2009). Seven guidelines for policy-making in an uncertain world. In D. Swanson & S. Bhadwal (Eds.), *Creating adaptive policies: A guide for policy-making in an uncertain world* (pp. 12–25). New Delhi, India: SAGE Publications India Pvt. Ltd. Retrieved from http://dx.doi.org/10.4135/9788132108245.n2

World Bank. (2013). *2nd Ethiopia economic update: Laying the foundations for achieving middle income status.* Washington, DC: Author. Retrieved from http://www-wds.worldbank.org/external/default/WDSContentServer/WDSP/IB/2013/12/03/0004 42464_20131203113831/Rendered/PDF/785010Revised00Box0379884B00PUBLIC0 .pdf)

World Resources Institute. (2009). *The national adaptive capacity framework: Key institutional functions for a changing climate.* Washington, DC: Author. Retrieved from http://pdf.wri.org/working_papers/NAC_framework_2009–12.pdf

REBECCA ADLER *works for LTSI, United Kingdom, as senior consultant, and is currently Project Manager and M&E consultant for the M&E of Ethiopian DFID climate investment programs, and also provides M&E support for the DANIDA Evaluation of Denmark's Climate Change funding for developing countries and the DFID Climate Public Private Partnerships Program.*

KIRSTY WILSON *is a principal consultant of LTSI, United Kingdom, and has been engaged in the DFID Ethiopia climate investment programs and as Team Leader for the Enhancing Community Resilience Program in Malawi.*

PATRICK ABBOT is LTSI, United Kingdom, managing director and involved with a range of M&E related assignments across the climate change and environment portfolio.

URSULA BLACKSHAW is a freelance consultant with over 30 years' experience in advising on organizational restructuring, institutional development and strategic planning processes in the public, private and civil sectors.

NEW DIRECTIONS FOR EVALUATION • DOI: 10.1002/ev

Karani, I., Mayhew, J., & Anderson, S. (2015). Tracking adaptation and measuring development in Isiolo County, Kenya. In D. Bours, C. McGinn, & P. Pringle (Eds.), *Monitoring
and evaluation of climate change adaptation: A review of the landscape. New Directions for
Evaluation, 147,* 75–87.

5

Tracking Adaptation and Measuring Development in Isiolo County, Kenya

Irene Karani, John Mayhew, Simon Anderson

Abstract

*This article highlights the utility of Tracking Adaptation and Measuring Devel-
opment (TAMD) in Isiolo County, Kenya. TAMD is a new adaptation monitoring
and evaluation (M&E) framework that was tested in Kenya for the first time
from April 2013 to March 2014. The article outlines the experiences of testing
the feasibility of the framework and its use in assisting policy makers at sub-
national level in adaptation planning, monitoring, and evaluation. It draws out
the experiences and lessons of using an ex ante approach in adaptation M&E in
a situation where adaptation actions are in the planning phase, using a scorecard
and theories of change to measure climate risk management processes, and potential adaptation/development benefits. It also highlights how climate change
adaptation M&E approaches can be simplified in order to get buy-in from
sub-national governments and communities so as to mainstream M&E in planning approaches for sustainability in developing countries.* © 2015 Wiley Periodicals, Inc., and the American Evaluation Association.

T racking Adaptation and Measuring Development (TAMD) was developed by Brooks et al. (2013) and offers a framework for use in
many contexts and at many scales to assess and compare the effectiveness of interventions that directly or indirectly assist populations
in adapting to climate change. According to Brooks et al. (2011), the
TAMD framework differs from those of the Pilot Programme for Climate

Figure 5.1. TAMD Framework

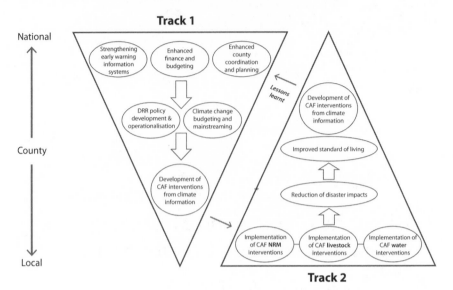

Resilience and the Adaptation Fund (cf. Climate Investment Funds [CIF], 2012; Adaptation Fund, 2011) combining top-down institutional indicators with bottom-up vulnerability indicators (p. 30). It also acknowledges and addresses problems associated with changing climatic baselines. TAMD provides an explicit framework for two paths, or "tracks"; Track 1 entails *assessing the capacity* of institutions to undertake effective climate risk management (CRM) actions (also called top-down), and Track 2 entails *assessing impacts* of interventions aimed at reducing vulnerability and the extent to which such interventions keep development on track (development performance or bottom-up). See Figure 5.1, adapted from Brooks and Fisher (2014).

TAMD testing was conducted simultaneously in six countries. Nepal, Ethiopia, and Pakistan used the framework in an ex post approach, evaluating completed projects that were deemed to be enhancing resilience irrespective of whether they were explicitly designed to do so. Mozambique, Cambodia, and Kenya attempted to enhance adaptation planning and M&E with the use of ex ante approaches (Pradhan, 2014).

Isiolo is one of Kenya's 47 counties, with pastoralism being the dominant livelihood in this semi-arid area. Over the years, its communities have continued to feel the impacts of climate variability through the increasing frequency of drought episodes and their negative impacts (Republic of Kenya, 2013). The county was chosen for the TAMD feasibility testing as it was the first to receive climate financing from the Department of International Development (DFID) for the establishment of a County Adaptation Fund (CAF). The objective of the CAF is to finance public-good investments for improved resilience to climate change through the County and Ward Adaptation Planning Committees (CAPCs and WAPCs, respectively) and the Adaptation Consortium.

TAMD was designed as an evaluative framework to be implemented ex post (after implementation). However, the researchers found that in Isiolo County the adaptation interventions and CRM processes were still in the planning phase; therefore, the testing of the framework took an ex ante approach (before implementation). It was also felt that for an effective ex post evaluation, a sound system that generates evidence on climate trends and adaptation benefits needed to be in place first.

This article presents the methodology, results, and lessons learned from testing the framework, in doing so informing adaptation M&E approaches in other developing countries and highlighting the advantages of mainstreaming M&E in planning processes.

Methodology

The research team used the TAMD checklist provided by Brooks et al. (2013, p. 27) as the general methodology. A scorecard for top-down processes and theories of change for bottom-up processes were the main tools utilized in designing an adaptation M&E system for Isiolo County. Before the TAMD tools were employed, the TAMD approach and key M&E concepts were simplified, for example, by using examples of how people conduct monitoring and evaluation in their daily lives and brainstorming sessions on the type of enabling environment that can allow reduction of vulnerability at community levels. The use of the term *TAMD* was also minimized so that it was seen as a methodology, as opposed to it being a project. Thus, the terms *monitoring of adaptation/development actions* were used to replace TAMD. Other terms that were simplified in the local languages included *outputs, outcomes, impacts, indicators,* and *resilience.*

Top-Down (Track 1) Process

For Track 1, county technical officers from the departments responsible for water, livestock, natural resource management, meteorology, planning and the National Drought Management Authority (NDMA) were brought together to identify and prioritize CRM activities required to build adaptive capacity at the community level. These activities were screened from the NDMA strategic plan, the draft Isiolo County Integrated Development Plan (ICIDP), and sectoral plans of the county.

The technical team assessed CRM processes through the use of Brooks' scorecard (Brooks et al., 2013, pp. 30–34). The scorecard measures CRM indicators in Track 1 through eight parameters, namely, climate change mainstreaming/integration into planning, institutional coordination, budgeting and finance, institutional knowledge/capacity, use of climate information, planning under uncertainty, participation, and awareness among stakeholders. Under each parameter, five questions need to be answered before scores are assigned. The type of scoring is chosen by stakeholders in terms of weighting (0–4) or percentages. In Isiolo County, percentages were used to depict the extent to which progress against the indicator was being made. If the scores were low, this information could be used to design interventions that enhance CRM processes and also to track subsequent progress. If the score was high, interventions would be designed to ensure that progress was maintained. From this scorecard, the county government prioritized, strengthening early warning systems, county budgeting and planning, and county coordination and planning.

Bottom-Up (Track 2) Process

Before they were supported in the development of theories of change (ToCs) per ward, it was important that communities defined the term *resilience* in their own contexts so as to understand how their planned adaptation actions might contribute to resilience. The researchers worked with six WAPCs to identify 20 ward adaptation/development interventions covering the water, livestock, and natural resource governance sectors that were in planning phases. Each of these wards was then assisted in developing their own specific ToC.

Linking Track 1 and Track 2

After the first top-down and bottom-up processes were completed, a composite theory of change was then developed by the county technical team and the WAPCs. This ToC linked the prioritized county CRM interventions identified through the scorecard process with the six ward ToCs as shown in Figure 5.2, adapted from Karani, Kariuki, and Osman (2014).

The methodology used above sought to learn lessons from three questions, namely:

1. How effective is using an ex ante approach versus an ex post approach in adaptation M&E?
2. To what extent can participatory processes be used in designing a ToC that links CRM activities (Track 1) with development outcomes (Track 2)?
3. How can the framework be used to inform planning at sub-national and community levels?

When the composite ToC was developed and expected changes and indicators were identified in the top-down, bottom-up, Track 1 and 2 linkage

Figure 5.2. Isiolo County Composite Theory of Change

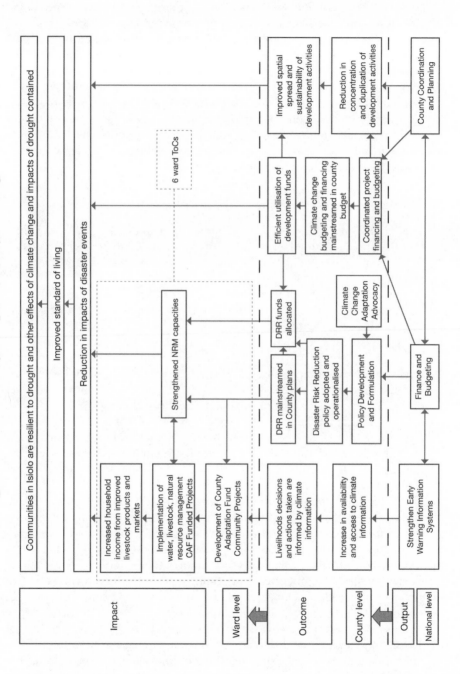

processes, the County Planning Unit proceeded to integrate relevant CRM and adaptation actions into the Isiolo County Integrated Development Plan (ICIDP) in order to mainstream adaptation planning and M&E.

The use of a participatory approach in testing the feasibility of TAMD was chosen, as it sought to enhance ownership of the data collected, the analysis, and the dissemination of lessons learned. It also sought to build the evaluative capacity among stakeholders, an approach also supported by Preskill (2009) and Preskill and Boyle (2008). This was achieved by examining various climate change and M&E definitions and processes with the county officials and WAPCs. For example, climate variability, maladaptation, outputs, outcomes, impacts, indicators, evaluation, and assumptions, before ToCs and M&E plans were developed.

Results

The results presented in Table 5.1 show the extent to which the TAMD steps were implemented with the use of the checklist in Brooks et al. (2013, p. 27), and the scorecard results are presented in Table 5.2, adapted from Karani, Kariuki, and Osman (2014).

Developing outputs, outcomes, and impacts and their indicators can be challenging, especially when engaging with people who do not have experience in M&E (Lennie, Tacchi, Koirala, Wilmore, & Skuse, 2011). Thus using phrases such as "signs of progress of change" for indicators was more easily understood. Using immediate, medium-term, and long-term changes in Kiswahili and Borana (local languages) also seemed to work for the definition of outputs, outcomes, and impacts. As a result, the major outcome of this process, which served as an eye-opener to the researchers, was the development of a composite ToC that combined CRM and adaptation interventions to define a common resilience impact statement agreed upon by both county officials and community representatives. This showed that they both understood how CRM and planned adaptation/development actions could enhance resilience collectively.

In addition, the use of the scorecard adapted from Brooks et al. (2013) proved to be a challenge, especially when explaining the difference between the scorecard indicators and the CRM indicators framed under the composite ToC that the end users had developed themselves. This challenge was overcome by rephrasing the scorecard as a tool to measure institutional capacities on CRM processes that could be used for monitoring adaptive capacity, alongside tracking progress against CRM indicators (Karani, I., Kariuki, N., & Osman, F., 2014).

Lessons from the Kenya study were used to influence the testing of TAMD in Mozambique, Ethiopia, Uganda, and Tanzania, despite different operating contexts. Challenges identified in Kenya led to changes in testing methodology. For example, the emphasis on collecting climate data from the beginning was emphasized in Mozambique (Artur, Karani, Gomes, Maló, &

New Directions for Evaluation • DOI: 10.1002/ev

Table 5.1. Results From Using the TAMD Steps

Step no.	Step description	Result
1.	Define the evaluation context and purpose (ex-post)	It was not possible to define the evaluation context ex post as adaptation interventions were in the planning phase. The understanding and design of specific adaptation interventions in Kenya is a recent phenomenon and only gained momentum during the national climate change planning process. Thus it was difficult to find completed adaptation actions, hence the change to an ex ante M&E planning process.
2.	Establish a theory of change (ToC)	It was found that although developing specific ward ToCs at the community level was possible, there was need to develop a composite one at county level in order to use development performance as a result of CRM processes (see Figure 5.2).
3.	Identify the relevant scales (global, national, regional, local)	It would be difficult to aggregate adaptation benefits at the national level if evidence is not generated at the lower levels because of a lack of M&E systems—hence the decision by the researchers to focus at sub-national and community levels.
4.	Locate outputs, outcomes, and impacts on the TAMD framework	Results at different levels were mapped onto the TAMD framework. This assisted in interrogating and strengthening the logic of how resilience is built through CRM and adaptation/development interventions.
5.	Identify the type of indicators required	CRM and adaptation outcome and impact indicators were developed. Each ward developed adaptation/development indicators as proxy measurements of resilience. The county government developed process indicators to measure CRM processes.
6.	Define the indicators	Scorecard indicators and bespoke indicators were defined for CRM processes, and qualitative and some quantitative indicators were used for development outcomes. These indicators were similar to development indicators used by the county government. They are presented in Brooks and Fisher (2014, p. 81).
7.	Gather data	After indicator definition, baseline data on CRM and adaptation actions was collected to the extent possible, as the interventions were yet to begin. This would have been extremely challenging in an ex post situation.
8.	Analyze indicators and data at different levels of Tracks 1 and 2 for attribution	This was not possible as there were no outcomes yet. In an ex ante M&E planning approach, opportunities for collecting adaptation monitoring data in order to address attribution in the long term are created.
9.	Address attribution	

(Continued)

Table 5.1. Continued

Step no.	Step description	Result
10.	Disseminate lessons from M&E so that interventions can be modified where necessary, and future interventions can be informed by these lessons	Lessons of using the ex ante approach from Kenya were crucial in shaping decisions by research teams in other countries (Mozambique, Uganda, Ethiopia, Tanzania), where the design of CRM and adaptation interventions were still in their planning phases. They were also used to inform the TAMD step by step guide by Brooks and Fisher (2014).

Table 5.2. Scorecard Results

CRM parameter	% score	Reasons
1. Extent to which climate change planning is integrated in county policies or processes	20	Isiolo County does not have a climate change strategy and there is limited expertise in climate change screening of development interventions
2. Extent to which there is institutional coordination of climate change interventions	85	Climate change adaptation interventions are coordinated across sectors by the county drought coordinator from NDMA.
3. Extent to which climate change financing is integrated into the county budget	55	The county had not yet budgeted or allocated finances for climate change. However CAF funding was available for adaptation activities at ward level.
4. Level of institutional climate change knowledge	65	The members of the CAPC had undergone climate change training but the knowledge of technical officers within the county government was still low.
5. Use of climate information	55	Some sectors of the county government (agriculture and water) took into account observational data and climate projections when planning. However, there was limited capacity to interpret and use climate information for scenario planning.
6. Planning under uncertainty	40	NDMA at county level updated its plans with climate information annually. However they did not use climate projections, nor did they consider maladaptation when planning.
7. Extent of participation during planning and decision-making processes in climate change adaptation	90	The design of ward adaptation actions took place after a highly participatory process, where women and other vulnerable groups had participated.
8. Level of climate change awareness amongst stakeholders	65	Only communities from 6 out of 10 wards in Isiolo County had been sensitized to climate change.

Anlaué, 2014). Mozambique went further and included this aspect in the national guidelines for local adaptation planning, which are now being implemented country-wide (Republic of Mozambique, 2014). In Uganda and Tanzania, TAMD training-of-trainers courses have included the aspect of building capacity in climate data collection and analysis (personal observation).

However, despite the progress achieved in testing the feasibility of the TAMD framework in an ex ante planning approach, there were challenges experienced regarding baseline data collection, which included inadequate technical and financial capacities to collect primary data and lack of cooperation from some key informants with secondary data. Collection of baseline climate trend data specific to Isiolo County was also a challenge because of the limited downscaling capacity of the meteorological unit in the county. Thus enhancing capacity for climate data collection and analysis is crucial for the success of an adaptation evaluation framework before its implementation.

Discussion

A change in approach from ex post evaluation to ex ante was seen as a more sustainable way of integrating M&E into planning processes. Ex post adaptation evaluations are costly; Chomitz (2010) proposed ex ante approaches for evaluation with regular updates of results. Stoorvogel, Claessens, Antle, Thornton, and Herrero (2011, p. 29) also emphasized that because data demands for ex post evaluations are high, the development of relatively simple methods for ex ante evaluation of adaptation at the household and system levels are therefore required. The results seen in Isiolo County show that ex ante approaches can be useful not only in enhancing climate change knowledge but general M&E appreciation among stakeholders where these capacities are low, also shown in the Mozambique case, in order to enhance sustainability of adaptation M&E (Artur et al., 2014, p. 6). Through using the ex ante approach, information on the benefits of adaptation can be used in an ex post evaluation together with climate trends to address the challenge of long time scales and possibly contribution/attribution.

There was an advantage of working with a county government and communities that had been sensitized on climate change issues for more than a year by NDMA, with vulnerability assessments also having been conducted before the TAMD approach was introduced. For communities or sub-national governments who have not been sensitized in climate change adaptation, the introduction of adaptation M&E frameworks such as TAMD may require more time and planning, as was the case in Mozambique (Artur et al., 2014, pp. 5–6).

It is important to establish a system that can use climate data in the normalization of adaptation indicators ex ante, if attribution is to be evidenced ex post. However by the end of the feasibility testing period, guidance on how this was to be done was still under discussion and was a major gap in the methodology and subsequent results. However, suggestions on how normalization can be done were finalized and published in December 2014 by Brooks and Fisher (2014, p. 54–59). The approach proposed will serve to guide other researchers and practitioners who want to apply TAMD in various contexts. However, there are also other methodologies of normalizing indicators proposed by the International Crops Research Institute for the Semi-Arid Tropics (International Crops Research Institute for the Semi-Arid Tropics [ICRISAT], 2009) in their vulnerability manual where they have used the United Nations Development Program's Human Development Index methodology to normalize indicators by identifying the functional relationship between the indicators and vulnerability.

Rai and Nash (2014) suggest that scorecards offer a relatively simple way to monitor institutional progress where key areas relevant to the intervention or desired outcomes have been identified. The United Nations Framework Convention on Climate Change is also exploring the use of scorecards in adaptation M&E (Adaptation Committee, 2014). Through the identification of weaknesses in CRM processes using the scorecard, the county technical team was able to identify and prioritize their interventions and proceed to integrate them into the County Integrated Plan. This type of mainstreaming is seen as an effective approach to climate change as it elicits benefits that include reducing policy conflicts, reducing risks and vulnerability. It has greater efficiency compared to managing adaptation separately, and can potentially leverage much larger financial flows in sectors affected by climate risks than the amounts available for financing adaptation separately (Lebel et al., 2012).

Climate financing for adaptation action can lead to enhanced motivation in M&E and learning by sub-national governments, as was the case in Isiolo County (Gesellschaft für Internationale Zusammenarbeit GmbH [GIZ] 2014; Karani., Anderson, S., & Kariuki, N., 2014) especially if information on adaptation indicators is similar to development indicators already present in the policy documents.

Conclusions

Returning to the initial questions that we sought to learn from in this study, researchers and adaptation practitioners can use ex ante planning approaches in evaluation when confronted with a situation where an ex post approach is not possible. A clear understanding of the adaptation M&E framework is required so that evaluation methodologies and tools are contextualized for the situation despite the challenges of baseline data collection and lack of climate information.

When adaptation M&E concepts are simplified, especially in contexts where climate change and M&E knowledge is low, the possibility of understanding how CRM interventions link to development outcomes and/or adaptation benefits is greater. Such outcomes are also enhanced where theories of change are developed through participatory processes that make use of the extensive experience of sub-national governments and resident communities in addressing extreme events and understanding what resilience is. If relevant stakeholders work towards creating composite adaptation ToCs together, it enhances coherent implementation and hopefully improves the collective impact of long-term interventions in increasing resilience. In turn, this can increase the uptake of adaptation M&E in capacity-building efforts.

TAMD, when used ex ante, is consistent with a Development Evaluation approach described by Patton (2010) as its application has to be contextualized, with innovation and real time monitoring in order to facilitate a continuous development cycle which ultimately will enhance development impacts and resilience. When institutionalized, the TAMD framework can be utilized in ex ante and ex post evaluation processes that will provide evidence of attribution between sub-nationally led CRM actions and development performance at grass-root levels in the longer term. This can be replicated at other sub-national levels of developing countries for result aggregation at national level, thus underscoring the importance of climate and development financing (Karani, Anderson, & Kariuki, 2014). Finally, theoretical adaptation M&E concepts require practical testing at different levels and continuous updating and revision with lessons learned if the uptake of these concepts is to become sustainable and useful to different end users.

References

Adaptation Committee. (2014). *Sixth meeting of the Adaptation Committee Bonn, Germany, 29 September–1 October 2014: Possible next steps and recommendations on monitoring and evaluation of adaptation* (Concept note). Bonn, Germany: UNFCCC Secretariat. Retrieved from http://unfccc.int/files/adaptation/application/pdf/ac6_-_draft_concept_note_me_8_sept_final.pdf

Adaptation Consortium. (2014). *Adaptation consortium news bulletin* [Online newsletter]. Retrieved from http://adaconsortium.org/images/publications/Briefing-Paper.pdf

Adaptation Fund. (2011). *Results framework and baseline guidance at project level*. Washington, DC: Author. Retrieved from http://adaptation-fund.org/sites/default/files/Results%20Framework%20and%20Baseline%20Guidance%20final%20compressed.pdf

Artur, L., Karani, I., Gomes, M., Maló, S., & Anlaué, S. (2014). *Tracking adaptation and measuring development in Mozambique*. London, United Kingdom: International Institute for Environment and Development (IIED). Retrieved from http://pubs.iied.org/10102IIED

Brooks, N., Anderson, S., Ayers, J., Burton, I., & Tellam, I. (2011). *Tracking adaptation and measuring development (TAMD)* (Climate Change Working Paper No. 1). London, United Kingdom: International Institute for Environmental Development (IIED). Retrieved from http://pubs.iied.org/pdfs/10031IIED.pdf

Brooks, N., Anderson, S., Burton, I., Fisher, S., Rai, N., & Tellam, I. (2013). *An operational framework for tracking adaptation and measuring development* (Climate Change Working Paper No. 5). London, United Kingdom: International Institute for Environmental Development (IIED). Retrieved from http://pubs.iied.org/pdfs/10038IIED.pdf

Brooks, N., & Fisher, S. (2014). *Tracking adaptation and measuring development (TAMD): A step-by-step guide* [Toolkit]. London, United Kingdom: International Institute for Environmental Development (IIED). Retrieved from http://pubs.iied.org/10100IIED

Chomitz, K. (2010). *Evaluating climate change adaptation efforts: Notes toward a framework.* Washington, DC: Independent Evaluation Group (IEG), World Bank. Retrieved from http://www.oecd.org/environment/cc/46220558.pdf

Climate Investment Funds (CIF). (2012). *Revised PPCR results framework.* Washington, DC: CIF Administrative Unit. Retrieved from https://www.climateinvestmentfunds.org/cif/sites/climateinvestmentfunds.org/files/Revised_PPCR_Results_Framework.pdf

Gesellschaft für Internationale Zusammenarbeit (GIZ). (2014). *Monitoring and evaluating adaptation at aggregated levels: A comparative analysis of ten systems.* Eschborn, Germany: Author. Retrieved from http://www.seachangecop.org/sites/default/files/documents/GIZ_2014-Comparative%20analysis%20of%20national%20adaptation%20M%26E.pdf

International Crops Research Institute for the Semi-Arid Tropics (ICRISAT). (2009). *Vulnerability assessment manual.* Hyderabad, India: Author. Retrieved from http://www.icrisat.org/what-we-do/mip/training-cc/october-2--3--2009/vulnerability-analysis-manual.pdf

Karani, I., Anderson, S., & Kariuki, N. (2014). *Institutionalizing adaptation monitoring and evaluation frameworks: Kenya* [Briefing paper]. London, United Kingdom: International Institute for Environmental Development (IIED), and Kenya, Nairobi: LTS Africa. Retrieved from http://pubs.iied.org/17251IIED.html?c=climate

Karani, I., Kariuki, N., & Osman, F. (2014). *Tracking adaptation and measuring development. Kenya research report.* London, United Kingdom: International Institute for Environmental Development (IIED). Retrieved from http://pubs.iied.org/10101IIED.html

Lebel, L., Li, L., Krittasudthacheewa, C., Juntopas, M., Vijitpan, T., Uchiyama, T., & Krawanchid, D. (2012). *Mainstreaming climate change adaptation into development planning.* Bangkok, Thailand: Stockholm Environment Institute Asia Office, and the Adaptation Knowledge Platform (AKP). Retrieved from http://ipcc-wg2.gov/njlite_download2.php?id=10827

Lennie, J., Tacchi, J., Koirala, B., Wilmore, M., & Skuse, A. (2011). *Equal access participatory monitoring and evaluation toolkit.* San Francisco, CA: Equal Access International. Retrieved from http://betterevaluation.org/toolkits/equal_access_participatory_monitoring

Patton, M. (2010). *Developmental evaluation applying complexity concepts to enhance innovation and use.* New York, NY: Guilford Press.

Pradhan, S. (2014). *TAMD multi-country meeting. Summary report.* London, United Kingdom: International Institute for Environmental Development (IIED). Retrieved from http://pubs.iied.org/G03808.html?c=climate

Preskill, H. (2009). Reflections on the dilemmas of conducting evaluations. In N. Birnbaum & P. Mickwitz (Eds.), *Environmental program and policy evaluation: Addressing methodological challenges. New Directions for Evaluation, 122,* 97–103.

Preskill, H., & Boyle, S. (2008). A conceptual model of evaluation capacity building: A multidisciplinary perspective. *American Journal of Evaluation, 29*(4), 443–459.

Rai, N., & Nash, E. (2014). *Evaluating institutional responses to climate change in different contexts* [Briefing paper]. London, United Kingdom: International Institute for Environmental Development (IIED). Retrieved from http://pubs.iied.org/17271IIED.html

Republic of Kenya. (2013). *Isiolo county: First county integrated development plan (2013–2017).* Kenya, Nairobi: Government of the Republic of Kenya.

Republic of Mozambique. (2014). *Methodological guidelines for the preparation of local adaptation plans*. Maputo, Mozambique: Ministry for Coordination of Environmental Affairs.

Stoorvogel, J. J., Claessens, L. F. G., Antle, J. M., Thornton, P. K., & Herrero, M. (2011). A novel methodology for ex ante assessment of climate change adaptation strategies: Examples from East Africa. In *Book of abstracts of the international conference on crop improvement: Ideotyping, and modelling for African cropping systems under climate change*. Stuttgart, Germany: University of Hohenheim.

Irene Karani is a director of LTS Africa, Kenya.

John Mayhew is a director of LTS International, United Kingdom.

Simon Anderson is the head of the climate change division at the International Institute of Environment and Development (IIED), United Kingdom.

Faulkner, L., Ayers, J., & Huq, S. (2015). Meaningful measurement for community-based adaptation. In D. Bours, C. McGinn, & P. Pringle (Eds.), *Monitoring and evaluation of climate change adaptation: A review of the landscape. New Directions for Evaluation, 147,* 89–104.

6

Meaningful Measurement for Community-Based Adaptation

Lucy Faulkner, Jessica Ayers, Saleemul Huq

Abstract

Evidence indicates ongoing tensions over effective climate change adaptation measurement. Focusing on community-based adaptation (CBA), we stress that some of these tensions stem from a lack of transparency around the knowledge and learning needs of different stakeholders engaged in CBA investments. Drawing on a participatory assessment of stakeholder information needs and appropriate scales required for effective monitoring and evaluation (M&E) for CBA, this article presents a new M&E for CBA framework. The framework identifies four levels at which M&E is to be undertaken by CBA practitioners and associated project stakeholders: participatory M&E at community level; M&E at individual project level and comparison across multiple project sites; M&E of capacity of institutions implementing CBA; and M&E of community of practice. The proposed framework tailors its M&E approaches according to these levels. By moving beyond the existing dominant donor-driven M&E perspective, we argue that this more nuanced approach enhances the usefulness of M&E by ensuring that the accountability of stakeholders engaged in CBA landscapes is legitimate across multiple scales. The framework is applicable for M&E of general development practice, as well as the climate change adaptation and resilience remit. © 2015 Wiley Periodicals, Inc., and the American Evaluation Association.

I ncreasingly complex and uncertain climatic risk landscapes mean it is not possible for people to possess complete knowledge of the kind of changes to anticipate or prepare for (Magis, 2010). It has therefore become critical for monitoring and evaluation (M&E) of climate change adaptation interventions to fulfill its potential in providing credible information that facilitates learning on climate change adaptation effectiveness (United Nations Framework Convention on Climate Change [UNFCCC], 2013; Villanueva, 2011); what works, what does not work, where, why, and, importantly, *for whom*—in addition to simply being a mechanism for measuring and reporting results (Bours, McGinn, & Pringle, 2014a).

This perspective is key to the climate change adaptation (CCA) arena known as community-based adaptation (CBA). CBA emerged in response to top-down approaches to adaptation planning and action that were criticized for failing to integrate adaptation and development in ways that address the diversity and complexity of local vulnerability contexts (Ayers & Huq, 2013). Accordingly, CBA follows an action research approach that operates at the community level. It identifies, assists, and implements community-based development activities that strengthen the capacity of the poorest and most marginalized people to adapt to climate change impacts (Ayers & Huq, 2013; Huq & Reid, 2007).

The rationale driving M&E for CBA should aim to reflect the central design principles upon which CBA is framed. Adaptation is understood as a process of change that builds on cultural norms facilitated from within (Ensor & Berger, 2009). It should therefore be undertaken by and not for communities (Huq, 2011). Consequently, M&E processes are to be participatory and empowering for engaged community stakeholders. This means that M&E processes should ask whether the definition of what successful CBA looks like from the community perspective is being assessed in legitimate ways that enhance learning and downward accountability (Chambers, 1997).

Yet a critical review of the current M&E for CBA discourse reveals barriers to achieving the intended deliberative institutional design of CBA. We recognize that much progress has been made in advancing both the theoretical underpinnings and practical applications of CBA (Ensor, 2014). This includes the identification of characteristics to measure local adaptive capacity (e.g., Jones, Ludi, & Levine, 2010). Similarly, progression has been shown in the development of participatory M&E (PM&E) tools and approaches to use at project level (e.g., Ayers, Anderson, Pradhan, & Rossing, 2012). However, we argue that there remains a lack of attention to and transparency around the knowledge and learning needs of different stakeholders engaged in CBA investments. These stakeholders operate from local to national and international level across horizontal and vertical scales. They are linked to a community through a CBA project and fundamentally influence M&E outcomes on the ground (Ensor, 2011, 2014). If M&E for CBA is to

NEW DIRECTIONS FOR EVALUATION • DOI: 10.1002/ev

provide critical support to the process of identifying what works, the question of who M&E works for demands greater consideration.

We propose that two distinct but related critiques surrounding M&E, CBA, and participatory approaches are relevant: (a) the assumption that local is always the appropriate scale for M&E for CBA (Dodman & Mitlin, 2011; Yates, 2014); and (b) the potential for powerful actors to define success, and by doing so shifting M&E for CBA away from the perspective of those most marginalized (Ensor, 2014).

In this article, we present a new M&E for CBA framework that responds to these critiques. The framework does not duplicate existing PM&E tools or methods for enabling CBA at the project level. Indeed some methods will fit the purpose of how to measure relevant tracks, or different stakeholder scales within the framework, as discussed below. Rather, the framework provides a multitrack approach that reframes M&E for CBA design. It provides a more comprehensive approach by addressing the diverse cross-scale information needs of CBA stakeholders and enables multidirectional knowledge and learning flows on effective adaptation. This is to ensure that accountability toward different stakeholders engaged in CBA investments is legitimate in multiple directions. The framework recommends indicators for assessing CBA effectiveness, and appropriate methodologies for undertaking M&E of climate change adaptation and resilience projects and programs.

The remainder of this article is divided into six sections. Section 2 discusses the above critiques in further detail. In response, section 3 presents the new M&E for CBA framework. Insights into early stages of framework operationalization follow in section 4. Framework challenges are presented in section 5, with conclusions and recommended next steps in sections 6 and 7 respectively.

Reframing M&E for CBA Design

As stated above, we identify the need to account better for different stakeholder information demands across scales when considering M&E for CBA practice. We recommend therefore that the community scale cannot be viewed in isolation (Ensor, 2014). Consequently, we propose that M&E for CBA must be seen within, and engage with, the wider political economy context.

We recognize that the notion of *community* is considered problematic (Norris, Stevens, Pfefferbaum, Wyche, & Pfefferbaum, 2008). We therefore advocate moving beyond normative assumptions that view the community as a fixed location, to a node in a network of multiscalar flows of relations, resources, and knowledge required to generate learning on adaptation (Yates, 2014, p. 18). This viewpoint provides a useful lens to rethink perceptions of appropriate stakeholders and scales to be acknowledged in M&E for CBA theory and practice.

However, the need to account better for multiple scales highlights the existing lack of consensus around *who* should define an intervention's "success." The vast literature on participation shows that facilitating PM&E in a truly meaningful way is rarely achieved because the tensions between the knowledge and learning needs of different stakeholders engaged in local CBA landscapes are not often effectively acknowledged (Cooke & Kothari, 2001; Estrella & Gaventa, 1998; Few, Brown, & Tompkins, 2007; Guijt, 1999, 2007).

This perspective is pertinent in light of the dominant donor-driven M&E approach in many developing-country contexts where CBA projects and programs operate (UNFCCC, 2013). Such an approach focuses on demonstrating value for money and results of CBA interventions for upward accountability purposes, in line with the information needs of funding agencies. In this remit, "successful" CBA is often defined by top-down institutional M&E processes.

Although the donor-driven M&E model will always be required, we conclude that it is not the only approach. We propose that a more enabling M&E agenda is needed for CBA that readdresses the balance of the existing "accountability gap" (Holland & Ruedin, 2012). This means an M&E agenda that promotes legitimate knowledge and learning flows on adaptation in multiple directions across scales to support more effective practice and empowerment at community level. This requires us as M&E practitioners to reconsider the purpose of M&E. We need to move beyond asking "Are we doing what we said we would do?" to "Does it work?", which begs the question of "Who does it work for?" Consequently, we assert that asking the question of "Who is this information for?" is critical as a first step in designing an approach to M&E for CBA, and M&E approaches for resilience and development at large. By asking this question, we stress that greater clarity can be brought to the subsequent key questions of "What are we measuring?" and "How do we do it?"

Conceptual Framework: M&E for CBA

Taking the above questions into consideration is central to the design of an emerging M&E framework designed for CBA, known as "ARCAB M&E for CBA." This framework was initially developed for a long-term action research program based in Bangladesh, with the International Centre for Climate Change and Development (ICCCAD), that aimed to generate longitudinal evidence on CBA effectiveness: Action Research for Community Adaptation in Bangladesh (ARCAB). The ARCAB program is a consortium of 11 international nongovernmental organizations (INGOs), and local, national and international research partners.

The ARCAB M&E for CBA framework aims to serve two purposes. First, M&E of CBA: generating rigorous scientific evidence on whether CBA improves the capacity of the poorest and most marginalized people to adapt to the impacts of climate change. Second, M&E for CBA: that is, using M&E as a mechanism for facilitating learning on how to respond to changing climate and vulnerability contexts. Both purposes also support a further objective: building a critical mass of best CBA practice that can be scaled up (i.e., integrated into subnational and national climate and development planning), and scaled out (i.e., targeted to reach a wider number of vulnerable people) in line with shifting CBA programming and funding architecture (Ayers & Abeysinghe, 2013; Huq & Faulkner, 2013; Rossing, Otzelberger, & Girot, 2014).

Framework design was undertaken in two stages in Bangladesh. First, the M&E information needs and accountability requirements of stakeholders engaged in CBA projects were identified by means of an iterative process consisting of four steps:

Step 1: Literature review, with a focus on existing M&E tools and approaches
Step 2: Stakeholder analysis, including different "layers" of communities, community-based organizations, NGOs, and local and national government and donors
Step 3: Field visits to explore the use of M&E tools and approaches in practice
Step 4: Stakeholder interviews, to elicit their M&E information needs and accountability requirements.

Research was undertaken in CBA projects operationalized by INGOs engaged in the ARCAB consortium. The results of this process are presented in Table 6.1.

Second, a hypothesis (Sidebar 1) was devised to articulate what effective CBA looks like based on the above stakeholder information needs. It provides a conceptual outline for what to measure to assess CBA effectiveness that guides indicator development across the framework. Correspondingly, this approach consists of two key outcome indicator areas. First, measuring the adaptive capacity and action of community stakeholders at project level in light of climate and other risks (known as downstream indicators). Second, measuring the capacity of relevant institutions to deliver adaptation support to community stakeholders and integrate local adaptation concerns into institutional processes from planning to implementation (known as upstream indicators). Further details on indicators are provided in Table 6.2 below.

Based on the above, the following multitrack framework for M&E for CBA is proposed (Figure 6.1).

Track 1: Participatory M&E (PM&E) at Community Level

The purpose of this track is for communities to drive PM&E processes through community monitoring groups actively. This aims to support the

Table 6.1. M&E Information Needs and Accountability Requirements of Stakeholders Engaged in CBA Projects and Programs

Stakeholders engaged in CBA project/ program landscapes (Identification of exact stakeholders in each group is intervention and context dependent)	Stakeholder information needs and accountability requirements (according to their own perspectives)
The poorest and most marginalized people vulnerable to climate change impacts. These stakeholders are the ultimate beneficiaries for whom CBA projects and programs are trying to affect change.	How well are CBA activities and interventions meeting our needs? How are institutions performing that claim to represent and support us? What are the effects of changes to climate and other risks that might influence our livelihoods and decision-making? What are our options for adaptation? How effective are proposed CBA strategies in light of climate change and other risks?
Local institutions. This includes both formal and informal institutions identified as relevant for enabling adaptation by vulnerable groups. This may include community-based organizations (CBOs), local NGOs, local government service delivery providers, and the private sector.	How are our activities supporting CBA? Are we promoting effective CBA processes and outcomes? How will climate change impacts affect what we are aiming to achieve?
INGOs/other institutions implementing CBA interventions. This includes organizations who support capacity building for those people most vulnerable to climate change impacts, and local institutions at project level.	How effective are we at delivering adaptation support? Are we translating our capacity into effective CBA on the ground? How can we draw information from our projects to measure change? How can we conduct a robust evaluation of results and impacts? How do we know if we are making a difference? How do we know if our CBA strategies are promoting "business as usual" rather than approaches that are considered more "transformative"? How can we improve our activities in the future?
The wider community of practice at national and international scales. This includes donors, researchers, and policy makers.	What is the value and effectiveness of CBA compared to other strategies? Why should we invest in CBA? What are the results of CBA investments? What really works for CBA and why? Which CBA approaches have the strongest impact in which contexts?

development of sustainable knowledge generation systems that empower communities and build adaptive capacity. In line with the information needs specified in Table 6.1, this track provides communities with a platform for multiple purposes—for example, assessing the progress of community adaptation plans and assessing the performance of organizations and service providers supporting CBA. Operationalizing this track can be guided by current PM&E approaches for CBA (e.g., Ayers et al., 2012).

SIDEBAR 1
ARCAB M&E for CBA hypothesis

Supporting effective CBA for the poorest and most marginalized people vulnerable to climate change impacts requires strengthening their knowledge and capacity to improve their long-term adaptive capacity in light of changes in climatic and other risks. It also simultaneously requires these stakeholders to have access to an enabling environment facilitating their ability to adapt. This requires local institutions to have the knowledge, capacity, and incentives to provide adaptation services and benefits to them. Together, therefore, these two components should result in evidence that people and institutions are actually adapting to climate change impacts through changing practice as a result of improved adaptive capacity and access to adaptation services.

Note: Adapted from Ayers and Faulkner (2012, p. 19).

Track 2: M&E of CBA at and Across Project Sites

This track assesses the progress of interventions in supporting effective CBA. To operationalize this track, longitudinal studies underpinned by a theory-of-change (ToC) approach are recommended to draw lessons from and across project sites. Information on why ToC is a stronger fit for M&E for CBA compared to a logical framework approach is provided in Bours, McGinn, and Pringle (2014b), and Faulkner (2013). As part of ToC design, indicators are to be identified based on the hypothesis in Sidebar 1. Context-specific indicators are then to be developed under each outcome indicator area as required for the project or program (Table 6.2). If a ToC approach is not currently used by an organization, the indicator areas presented here can be applied to fit existing M&E methodologies.

Track 3: M&E of Capacity of INGOs/Other Institutions Implementing CBA

This track assesses the capacity of INGOs and other organizations to deliver effective CBA interventions. It looks to answer the questions, "Where are we now to support effective CBA?" and "What do we need to do to improve our performance?" Using a ToC approach is recommended to provide a clear

Table 6.2. ARCAB M&E for CBA Indicators for Assessing CBA Effectiveness

Downstream indicators (assessing community stakeholders at project level)	Upstream indicators (assessing relevant institutions)
Indicators around adaptive capacity are challenging to define given the uncertainty surrounding the concept. However, it is widely agreed that good development coupled with access to and ability to use information related to climate risks, are prerequisites for CBA. Progress against context-specific development indicators in light of climate and other risks therefore provides one set of indicators for adaptive capacity. This includes indicators on poverty, asset bases and livelihoods, food security, health, and disaster risk reduction (DRR). Indicators of awareness and the ability to use climate information in adaptive decision-making is another. Indicators showing evidence of adaptive behaviors are also necessary in order to assess if people are actually adapting over time. This requires indicators, for example, around the shifting of livelihood strategies that promote adaptation rather than coping.	For effective CBA that is sustainable over time, community groups require access to (local) institutions that support an enabling environment for adaptation. This means context-specific indicators assessing institutional capacity and mainstreaming are required (Mainstreaming means integrating information and processes that aim to address climate change adaptation into ongoing institutional development planning and programming). Indicators therefore need to assess: Institutional and service accessibility and inclusiveness. Knowledge and capacity of (local) institutions to integrate climate risk management into existing planning and provision. Knowledge of climatic variability and climate change, including how these climatic risks manifest at the local scale and how it is likely to affect those most vulnerable to its impacts. The delivery of institutional adaptation support to most vulnerable groups.
Example indicators	*Example indicators*
Evidence of changes in value of assets/improved livelihood outcomes (in light of climate and other risks). Evidence of increased skills and resources to undertake new and improved practices. Number of vulnerable people using climate information in decision-making processes. Evidence of changing attitudes to risk taking and longer-term planning.	% of vulnerable groups (disaggregated by gender) actively participating in local institutional planning and budgeting meetings. % of annual institutional budget allocated to vulnerable group adaptation strategies. Level of knowledge of climate variability/potential climate change impacts. Evidence of ability to discuss, generate, and adapt existing practices to changing circumstances if required.

Note: Based on Ayers and Faulkner (2012, pp. 14, 20–23).

Figure 6.1. The ARCAB M&E for CBA framework

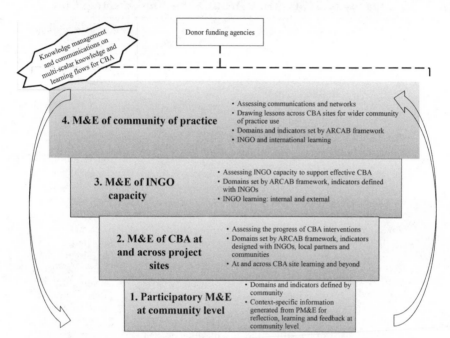

Note: Adapted from "Figure 12: A Multi-track Strategy for ARCAB M&E for CBA" from Ayers and Faulkner (2012, p. 18).

roadmap of how this is to be achieved. Appropriate indicators guided by the ARCAB methodology are then to be defined with the relevant stakeholders of the institution in question. It is useful to assess results using the ARCAB CBA Resilience Scale (Table 6.3).

The ARCAB CBA Resilience Scale moves horizontally from development to adaptation to climate variability, including disaster risk reduction, to adaptation to climate change. Vertically, the scale moves from business as usual approaches to development, to those considered transformative (i.e., using methods and approaches that promote change for sustainable CBA outcomes rather than those that maintain the status quo). Moving toward transformative approaches in all domains is recommended. Examples are provided in Table 6.3. For detailed information and practical examples of use, see Ayers and Faulkner (2012), Faulkner and Ali (2012), Reid, Faulkner, and Weiser (2013), and Huq and Faulkner (2013).

Track 4: M&E of Community of Practice

This track is yet to be developed. However, it aims to translate what is happening across scales into useful evidence that is responsive to the information needs of relevant stakeholders in Table 6.1.

Table 6.3. The ARCAB CBA Resilience Scale Used to Assess the Capacity of INGOs/Other Institutions Implementing CBA

	Development	Adaptation to climate variability	Adaptation to climate change
Business as usual	Projectized development Inflexible linear planning Poor participation Short-term focus	Use of scientific information on climate variability and disaster Largely disaster response than preparedness	Climate impacts focused Prioritizes climate impact information over local knowledge Top down approach rolled out rather than scaled out Mainly technological interventions implemented
Transformative	Empowerment of vulnerable households Community-driven Bottom up accountability Flexible responsive planning Strong institutional processes Good participatory approaches	Strong community knowledge of climate variability/ disaster impacts Use of scientific information on climate variabil-ity/disasters Development needs addressed as first step towards adaptation	New climate knowledge: blending climate change science with meaningful local knowledge Climate change adaptation mainstreamed across all operational project levels Scaling out driven by knowledge changes in stakeholder groups Long-term focus

Note: Adapted from "Figure 15: ARCAB's CBA Transformation Scale to Assess Action Partner Capacity" from Ayers and Faulkner (2012, p. 24).

ARCAB M&E for CBA Framework Application

Although the utilization of the ARCAB M&E for CBA framework is at an early stage, it has gained rapid attention, recognition, and uptake by the international community (e.g., Baldwin, Faulkner, Hawrylyshyn, Phelan, & Stone, 2014; Faulkner, 2013; Faulkner & Ali, 2012). To date, certain tracks of the framework have been operationalized in response to institutional demands requesting three key M&E objectives: (a) preintervention M&E design, (b) end-of-intervention evaluation, and (c) an assessment of CBA effectiveness to showcase how to scale up what works. As presented below, the framework has shown to be universal in application and use for stakeholders engaged in different livelihood zones and adaptation contexts.

NEW DIRECTIONS FOR EVALUATION • DOI: 10.1002/ev

Similarly, evidence shows that it translates to non-CCA–specific investments (e.g., Reid et al., 2013), plus those targeting CCA and resilience goals (e.g. Faulkner, 2012; Huq & Faulkner, 2013).

(1) Preintervention M&E Design

The framework has aided M&E strategy design for INGO CCA and resilience projects in Bangladesh, Myanmar, and Somaliland (Baldwin et al., 2014; Faulkner, 2012, 2013). This has included the development of a collaborative ToC linking cross-scale stakeholders to support longer-term programming approaches (Faulkner, 2012).

Evidence shows that the framework is effective in designing M&E systems to ensure that M&E is front-loaded for learning on CCA/resilience and reporting purposes (Baldwin et al., 2014); in enhancing existing M&E approaches already in place; and in guiding ongoing and future intervention design (Faulkner, 2012, 2013).

(2) End-of-Intervention Evaluation

Two components of the framework have been used to assess CBA (Faulkner & Ali, 2012), and participatory natural resource management interventions (Reid & Faulkner, 2015; Reid et al., 2013) for INGOs in Bangladesh and Ethiopia. First, indicators were identified to evaluate what role initiatives had in delivering adaptation support for project stakeholders. Second, findings were analyzed by adapting the ARCAB CBA Resilience Scale in track 3 to assess intervention results and INGO capacity. Institutional feedback on framework use specified its utility in generating useful reflections on community-based change processes and adaptation support that informed upgrading of interventions.

(3) Assessment of CBA Effectiveness to Scale Up What Works

The framework has proven beneficial in illustrating how effective CBA can be scaled up and out (Huq and Faulkner, 2013). Here, the Global Environment Facility (GEF) Small Grants Programme (SGP) funded CBA project in Namibia, implemented by the United Nations Development Programme (UNDP), was evaluated. Based on assessment results, context-specific ToCs were developed showcasing how effective outcomes could be taken to scale. Recommendations were also presented on shifts in existing institutional thinking processes and funding mechanisms to enhance adaptation strategy and impact further.

Framework Challenges

The framework calls for a reassessment of existing M&E for CBA practice. Consequently, an enabling institutional environment to support its operationalization may require changes in individual practitioner and wider institutional mindsets across scales. This includes the knowledge and use of ToC in the CBA remit. Capacity building of INGO practitioners in this domain remains in its early stages with not all organizations familiar with the approach. It is therefore likely to take time for ToC, and for this framework which utilizes it, to be adopted widely. Similarly, navigating considerations of the time and availability of resources to facilitate ToC and PM&E approaches in already time and resource-limited environments needs to be acknowledged. This refers to both implementing institutions and within communities themselves.

Moreover, digging deeper into the management of contested multistakeholder knowledge boundaries and power symmetries is required. The framework acknowledges these issues, yet they warrant greater attention as framework operationalization develops. This includes the call for genuine participation (Few et al., 2007). Also, given the framework's multiscalar makeup, due attention to its use is required to limit any potential trade-offs between M&E quality and scale (Rossing, Otzelberger, & Girot, 2014). This holds relevance for the design of the fourth track of the framework: M&E of community of practice. Challenges will include how to draw local lessons from across CBA sites effectively and translate them into generalizable insights for policy makers (Larsen, Swartling, Powell, Simonsson, & Osbeck, 2011).

Moreover, existing tensions surrounding the notion of community are relevant to the framework. We assert that questioning assumptions that presume cohesion and trust within communities is useful to support more effective practice (Cannon, 2008; Walker, Devine-Wright, Hunter, High, & Evans, 2010). Similarly, we highlight that greater awareness in defining who forms the community is valuable, given that it is often interchangeable (Jupp, 2013). Likewise, measurement challenges surrounding the unit of a community from the outset requires notice (Ayers, 2011).

Conclusion

We argue that M&E practitioners should consider the knowledge and learning needs of stakeholders engaged in CBA investments across scales for M&E to be effective. Correspondingly, this chapter presents a new multitrack framework that reframes M&E for CBA design. In doing so, it enhances the usefulness of M&E by ensuring that accountability to all stakeholders is legitimate in multiple directions. It is transparent about the diverse cross-scale information needs of different audiences, and hence tailors its M&E approaches accordingly. Consequently, we assert that this approach is relevant for both CCA/resilience specialists and nonspecialists alike.

Current framework applications address key M&E objectives: preintervention M&E design, end-of-intervention evaluation, and showcasing how to scale up what works. Evidence highlights framework applicability in diverse livelihood and adaptation contexts. Also, it is transferable to sustainable development investments, plus those targeting more specific CCA and resilience goals.

Framework challenges include potential changes in practitioner and institutional mindsets given a needed reassessment of M&E for CBA practice. This includes CBA practitioner capacity building in the knowledge and use of ToC. Similarly, M&E and CBA practitioners will be required to pay greater attention to managing contested multistakeholder knowledge boundaries and power symmetries across scales as framework development progresses. Likewise, it is recommended that they consider tensions surrounding the notion of community to enhance more effective practice. In addition, how framework users can effectively draw and translate local lessons from across different CBA sites into generalizable insights for policy makers is of concern.

Next Steps

The M&E for CBA framework presented in this chapter is to be considered a work in progress. We aim for it to be refined through practice by different users across relevant scales. Further user feedback on this approach to inform framework enhancement is expected to be shared at the Ninth International Conference on Community-based Adaptation in Kenya, in April 2015, with the theme of "Measuring and Enhancing the Effectiveness of Adaptation."

References

Ayers, J. (2011). Background paper prepared for the experts workshop on participatory monitoring and evaluation for community-based and local adaptation. 17–18 February 2011. London, United Kingdom: CARE UK.

Ayers, J., & Abeysinghe, A. (2013). International aid and adaptation to climate change. In R. Falkner (Ed.), *The handbook of global climate and environment policy*. Oxford, United Kingdom: Blackwell Publishing Ltd.

Ayers, J., Anderson, S., Pradhan, S., & Rossing, T. (2012). *Participatory monitoring, evaluation, reflection and learning for community-based adaptation: A manual for local practitioners*. London, United Kingdom: CARE International. Retrieved from http://www.care.org/sites/default/files/documents/CC-2012-CARE_PMERL_Manual_2012.pdf

Ayers, J., & Faulkner, L. (2012). *Action research for community-based adaptation (CBA) in Bangladesh (ARCAB): M&E and baseline strategy for CBA (Final report)*. Dhaka, Bangladesh: International Centre for Climate Change and Development (ICCCAD).

Ayers, J., & Huq, A. (2013). Adaptation, development and the community. In J. Palutikof, S. L. Boulter, A. J. Ash, M. Stafford Smith, M. Parry, M. Waschka, & D. Guitart (Eds.), *Climate adaptation futures*. Oxford, United Kingdom: John Wiley & Sons Ltd.

Baldwin, M., Faulkner, L., Hawrylyshyn, K., Phelan, J., & Stone, J. (2014). *Monitoring and evaluation framework and strategy for the BRACED Myanmar community resilience programme*. Yangon, Myanmar: BRACED Beda Alliance—Plan International, Action Aid, World Vision, UN Habitat, BBC Media Action and Myanmar Environment Institute.

Bours, D., McGinn, C., & Pringle, P. (2014a). *Guidance note 1: Twelve reasons why climate change adaptation M&E is challenging*. Phnom Penh, Cambodia: SEA Change CoP and Oxford, United Kingdom: UKCIP. Retrieved from http://www.ukcip.org.uk/wordpress/wp-content/PDFs/MandE-Guidance-Note1.pdf

Bours, D., McGinn, C., & Pringle, P. (2014b). *Guidance note 3: Theory of change approach to climate adaptation planning*. Phnom Penh, Cambodia: SEA Change CoP, and Oxford, United Kingdom: UKCIP. Retrieved from http://www.ukcip.org.uk/wordpress/wp-content/PDFs/MandE-Guidance-Note3.pdf

Cannon, T. (2008). *Reducing people's vulnerability to natural hazards: Communities and resilience* (Research Paper No. 2008/34). Helsinki, Finland: United Nations University (UNU-WIDER). Retrieved from http://www.wider.unu.edu/publications/working-papers/research-papers/2008/en_GB/rp2008--34/

Chambers, R. (1997). *Whose reality counts: Putting the first last*. London, United Kingdom: Intermediate Technology Publishing.

Cooke, B. and Kothari, U. (2001). *Participation: The new tyranny?* London: Zed Books

Dodman, D., & Mitlin, D. (2011). Challenges to community-based adaptation. *Journal of International Development, 25*(5), 640–659.

Ensor, J. (2011). *Uncertain futures: Adaptation development to a changing climate*. Rugby, United Kingdom: Practical Action Publishing.

Ensor, J. (2014). Emerging lessons for community-based adaptation. In J. Ensor, R. Berger, & S. Huq (Eds.), *Community-based adaptation to climate change: Emerging lessons*. Rubgy, United Kingdom: Practical Action Publishing.

Ensor, J., & Berger, R. (2009). Community-based adaptation and culture in theory and practice. In W. N. Adger, I. Lorenzoni, & K. L. O'Brien (Eds.), *Adapting to climate change: Thresholds, values, governance*. Cambridge, United Kingdom: Cambridge University Press.

Estrella, M., & Gaventa, J. (1998). *Who counts reality? Participatory monitoring and evaluation: A literature review* (Working Paper 70). Brighton, United Kingdom: Institute of Development Studies (IDS). Retrieved from https://www.ids.ac.uk/files/Wp70.pdf

Faulkner, L. (2013). *SmartFarm monitoring and evaluation framework and strategy* (WorldFish White Paper 2013–47). Penang, Malaysia: WorldFish. Retrieved from http://pubs.worldfishcenter.org/resource_centre/WF-2013-47.pdf

Faulkner, L. (2012). *Save the Children monitoring and evaluation framework for DRR/CCA in Odwayne district, Toghdeer region, Somaliland* (Technical Report). Somaliland, Somalia: Save the Children.

Faulkner, L., & Ali, I. (2012). *Moving towards transformed resilience: Assessing community-based adaptation in Bangladesh*. Dhaka, Bangladesh: ActionAid Bangladesh, International Centre for Climate Change and Development (ICCCAD). Retrieved from http://www.actionaidusa.org/bangladesh/publications/moving-towards-transformed-resilience-assessing-community-based-adaptation

Few, R., Brown, K., & Tompkins, E. (2007). Public participation and climate change adaptation: Avoiding the illusion of inclusion. *Climate Policy, 7*(1), 46–59.

Guijt, I. (1999). *Participatory monitoring and evaluation for natural resource management and research: Socio-economic methodologies for natural resources research*. Kent, United Kingdom: Natural Resources Institute. Retrieved from http://www.nri.org/projects/publications/bpg/bpg04.pdf

Guijt, I. (2007). *Assessing and learning for social change: A discussion paper*. Brighton, United Kingdom: Institute of Development Studies (IDS). Retrieved from http://www.ids.ac.uk/publication/assessing-and-learning-for-social-change-a-discussion-paper

Holland, J., & Ruedin, L. (2012). *Monitoring and evaluating empowerment processes*. Stockholm, Sweden: Swiss Agency for Development and Co-operation. Retrieved from http://www.oecd.org/dac/povertyreduction/50158246.pdf

Huq, S. (2011). Adaptation: Resources now to plan and implement. *Sustainable Development Opinion*, November 2011. London, United Kingdom: International Institute for Environment and Development (IIED). Retrieved from http://pubs.iied.org/pdfs/17117IIED.pdf

Huq, S., & Faulkner, L. (2013). *Taking effective community-based adaptation to scale: An assessment of the GEF Small Grants Programme (SGP) community-based adaptation (CBA) project in Namibia*. New York, NY: United Nations Development Programme (UNDP). Retrieved from http://icccad.net/wp-content/uploads/2014/05/cba_namibia_report_final_june_13_11am.pdf

Huq, S., & Reid, H. (2007). *Community-based adaptation: An IIED briefing*. London, United Kingdom: International Institute for Environment and Development (IIED). Retrieved from http://pubs.iied.org/pdfs/17005IIED.pdf

Jones, L., Ludi, E., & Levine, S. (2010). *Towards a characterisation of adaptive capacity: A framework for analysing adaptive capacity at the local level*. London, United Kingdom: Overseas Development Institute (ODI). Retrieved from http://www.odi.org/sites/odi.org.uk/files/odi-assets/publications-opinion-files/6353.pdf

Jupp, E. (2013). I feel more at home here than in my own community: Approaching the emotional geographies of neighbourhood policy. *Critical Social Policy, 33*, 532–553.

Larsen, R. K., Swartling, A. G., Powell, N., Simonsson, L., & Osbeck, M. (2011). *A framework for dialogue between local climate adaptation professionals and policy makers. Lessons from case studies in Sweden, Canada and Indonesia*. Stockholm, Sweden: Stockholm Environment Institute. Retrieved from http://www.sei-international.org/mediamanager/documents/Publications/SEI-ResearchReport-Larsen-AFrameworkForDialogueBetweenLocalClimateAdaptationProfessionalsAndPolicyMakers-2011.pdf

Magis, K. (2010). Community resilience: An indicator of social sustainability. *Society & Natural Resources. An International Journal, 23*(5), 401–416.

Norris, F. H., Stevens, S. P., Pfefferbaum, B., Wyche, K. F., & Pfefferbaum, R. L. (2008). Community resilience as a metaphor, theory, set of capacities, and strategy for disaster readiness. *American Journal of Community Psychology, 41*(1–2), 127–150.

Reid, H., & Faulkner, L. (2015). Assessing how participatory, community-based natural resource management initiatives contribute to climate change adaptation in Ethiopia. In W. Leal Filho (Ed.), *Handbook of climate change adaptation, Volume III: Climate change adaptation, agriculture and water security*. New York, NY: Springer.

Reid, H., Faulkner, L., & Weiser, A. (2013). *The role of community-based natural resource management in climate change adaptation in Ethiopia* (Climate Change Working Paper No. 6). London, United Kingdom: International Institute for Environment and Development (IIED). Retrieved from http://pubs.iied.org/pdfs/10048IIED.pdf

Rossing, T., Otzelberger, A., & Girot, P. (2014). Scaling-up the use of tools for community-based adaptation: Issues and challenges. In J. Ayers, H. Reid, L. Schipper, S. Huq, & A. Rahman (Eds.), *Community-based adaptation to climate change: Scaling it up*. London, United Kingdom: Routledge.

United Nations Framework Convention on Climate Change. (2013). Workshop on the monitoring and evaluation of adaptation, Nadi, Fiji, 9–11 September 2013 [Background note]. Bonn, Germany: Author. http://unfccc.int/files/adaptation/cancun_adaptation_framework/adaptation_committee/application/pdf/ac_m&re_ws_background_note_16august2013.pdf

Villanueva, P. S. (2011). *Learning to ADAPT: Monitoring and evaluation approaches in climate change adaptation and disaster risk reduction—Challenges, gaps and ways forward* (Strengthening Climate Resilience Discussion Paper No. 9). Brighton, United Kingdom: Institute of Development Studies (IDS). Retrieved from http://www.ids.ac.uk/files/dmfile/SilvaVillanueva_2012_Learning-to-ADAPTDP92.pdf

Walker, G., Devine-Wright, P., Hunter, S., High, H., & Evans, B. (2010). Trust and community: Exploring the meanings, contexts and dynamics of community renewable energy. *Energy Policy*, *38*(6), 2655–2663.

Yates, J. S. (2014). Power and politics in the governance of community-based adaptation. In J. Ensor, R. Berger, & S. Huq (Eds.), *Community-based adaptation to climate change: Emerging lessons*. Rubgy, United Kingdom: Practical Action Publishing.

LUCY FAULKNER *is an independent consultant and researcher who was engaged in the design of the ARCAB M&E for CBA framework and has led its operationalization. In doing so, she has collaborated with the International Centre for Climate Change and Development (ICCCAD), Bangladesh.*

JESSICA AYERS *is a senior policy adviser at the UK Department for Energy and Climate Change. At the time of writing, she worked as a researcher for the International Institute for Environment and Development (IIED), United Kingdom, on pro-poor climate change planning, evaluation and finance in South Asia and Africa.*

SALEEMUL HUQ *is the director of the International Centre for Climate Change and Development (ICCCAD).*

7

What Indicates Improved Resilience to Climate Change? A Learning and Evaluative Process Developed From a Child-Centered, Community-Based Project in the Philippines

Joanne Chong, Anna Gero, Pia Treichel

Abstract

Community-based climate change adaptation and resilience (CCAR) projects increasingly recognize that climate change impacts are localized, requiring context-specific interventions. Conventional approaches to monitoring and evaluation (M&E) are, however, ill-suited to understanding the impact of such CCAR interventions. To address this gap, research based on a child-centered community-based adaptation project in the Philippines has developed a practical and replicable process for developing evidence-based, local-level indicators of effective adaptation. The process assesses how the project influenced children's knowledge, advocacy efforts, and impact on policy and practice. Evidence was generated from qualitative inquiry, primarily through focus group discussions with children. The analysis included scalar ratings to help to meet quantitative reporting requirements. A detailed guide was developed for implementing agencies to systematically understand, measure, and communicate evidence. The process can also be translated to community development projects seeking to evaluate change under uncertainty. © 2015 Wiley Periodicals, Inc., and the American Evaluation Association.

NEW DIRECTIONS FOR EVALUATION, no. 147, Fall 2015   © 2015 Wiley Periodicals, Inc., and the American Evaluation Association.   Published online in Wiley Online Library (wileyonlinelibrary.com) • DOI: 10.1002/ev.20134

Climate change adaptation and resilience (CCAR) requires integrated action across actors and scales. However, impacts and vulnerabilities are ultimately specific to local contexts, livelihoods are often directly dependent on local environments, and communities are at the front line of responding to the impact of climate change, particularly in the context of climate-related disasters. Recognizing that adaptation must occur in communities and reflect community priorities (Measham et al., 2011), donors and practitioners are increasingly funding and implementing community-based CCAR projects. Such programming also needs to recognize that climate change has particular impacts on children and youth (UNICEF Office of Research, 2014); that young people have the potential to support adaptation efforts as members of their families, schools, and communities into the future as tomorrow's leaders (Back, Cameron, & Tanner, 2009); and that a child-centered approach to community-based CCAR can benefit the entire community (Mitchell & Borchard, 2014).

The Child-Centered Community-Based Climate Change Adaptation (CCCBA) project in the Philippines, supported by the Australian government, is being implemented by Plan International, Save the Children, and the Institute for Sustainable Futures of the University of Technology Sydney. Plan International's child-centered approach to development recognizes that children are among the most vulnerable to the impacts of climate change and also acknowledges children's potential as agents of change.

Our research developed a replicable process, based on the project's implementation and focus-group discussions (FGDs) with children, for developing evidence-based, local-level indicators of effective adaptation. It acknowledges that conventional monitoring and evaluation (M&E) approaches are ill-suited to understanding the impact of CCAR interventions (Intergovernmental Panel on Climate Change [IPCC], 2014), due to the complex, nonlinear, multifaceted, localized, and uncertain nature of CCAR (Arnold, Mearns, Oshima, & Prasad, 2014; Bours, McGinn, & Pringle, 2014b). Several studies into climate change and children focus on quantifying specific impacts—for example, undernutrition resulting from decreased crop yields (Lloyd, Kovats, & Chalabi, 2011). Our research investigated a wider spectrum of issues relevant to children and climate change, including considering children's perceptions of risk to climate change (Tanner, 2010) and child-centered responses to climate change (Tanner et al., 2010).

About the CCCBA Project

The Child-Centered Community-Based Climate Change Adaptation (CCCBA) project (2012–2015) was implemented in 40 barangays (the smallest administrative division in the Philippines) across four provinces (Aurora, Eastern Samar, Northern Samar, and Southern Leyte). The Philippines was ranked second on the 2014 World Risk Index (United Nations

University—Institute for Environment and Human Security [UNU-EHS], 2014) and 117th out of 187 countries on the United Nations Human Development Index (2013 data, see United Nations Development Programme [UNDP], 2014). It is vulnerable to climate change impacts including changing rainfall patterns and temperatures, rising sea levels, and increased extreme weather events such as typhoons and increased storm surges (Philippine Atmospheric, Geophysical and Astronomical Services Administration [PAGASA], 2011). Climate change inhibits children's ability to access their full range of rights, compromises access to education, and poses health risks including heat stroke, diarrhea, vector-borne diseases, and malnutrition (Perera, 2008; Sheffield & Landrigan, 2011). Children are disproportionately affected by disasters in terms of loss of life and health impacts, and rates of abuse typically increase in the aftermath of natural disasters (Child Protection Working Group [CPWG], 2012; Risdell & McCormick, 2013).

The CCCBA project's theory of change (ToC) is based on three interlinked pillars: knowledge (increased knowledge about CCAR), advocacy (mobilizing and equipping children to influence community and government), and local policy and practice (strengthening the capacity of children and communities to participate in adaptation planning and actions). Project activities include school curricula development, community education, participatory climate vulnerability and capacity assessments (CVCAs), locally developed and implemented adaptation initiatives, training on using multimedia for communication and advocacy, support for out-of-school youth groups, and support for local government leaders to incorporate climate change into plans and policies. As there is no single mode of child participation appropriate for tackling CCAR in all instances (Landsdown, 2006; Shier, 2001; Tanner, 2010), the CCCBA implements various approaches to engage children and their communities on CCAR.

Background to Our Approach: Linking Focus Groups and Indicators

In line with the assertion that "[c]ommunity driven development programs could provide an important laboratory for studying the indicators and impacts of resilience-building efforts" (Arnold et al., 2014, p. 25), our research developed indicators using the CCCBA project as an evidence base, tested through an iterative process of learning and redesign. Several studies document CCAR indicators about changes that occur in the community, including some that specifically relate to children. For example, "people have technical skills to implement adaptation strategies" (CARE, 2010, p. 6), and "children participate in vulnerability and adaptation assessments with their community on a regular basis" (Hawrylyshyn, Elegado, & Tribunalo, 2011, p. 7). Our research develops and documents the process for conducting and analyzing FGDs to generate evidence against community-based indicators.

New Directions for Evaluation • DOI: 10.1002/ev

An indicator framework was selected to enable the implementing organizations to understand, measure, and communicate outcomes systematically (Lamhauge, Lanzi, & Agrawala, 2012). A set of indicators also addresses the complexity of CCAR that means there is no single metric for adaptation (Bours, McGinn, & Pringle, 2014a; Brooks, Anderson, Ayers, Burton, & Tellam, 2011). Our indicator set is based on the project's ToC and focuses on qualitative change indicators, in contrast to M&E frameworks that typically rely heavily on quantitative indicators of inputs and outputs—such as number of children immunized, or attending school.

FGDs were the primary method of inquiry, supplemented by additional semi-structured interviews. FGDs are widely used and recognized as an effective qualitative method for evaluation research (Patton, 2015). Qualitative methods reflect well-established research and practice on participatory M&E and community development and on the value of directly involving those young people who are project beneficiaries in monitoring and evaluation activities (Department for International Development [DFID], 2010).

The potential limitations of the focus-group method are well documented (Stewart & Shamdasani, 2015). One common criticism is that details are rarely provided for analyzing and interpreting FGD data (Massey, 2011); our approach, in contrast, is built around developing and documenting a detailed process for gathering and analyzing evidence from FGDs. We also conducted most of the focus groups in schools, addressing the commonly cited disadvantage that FGDs are not typically conducted in locations where social interactions normally occur (Patton, 2015). FGDs are a powerful means of inquiry and an effective approach to target evidence gathering given the context (climate change and children) and goals (to facilitate learning for project implementers).

Designing the Indicators and Focus Groups

We began with a review of literature and consultation with project staff to translate the ToC into key changes that we might expect to see in a community that has become more resilient, to form the basis of a preliminary draft of the indicators. These indicators were field tested and developed via several rounds of FGDs.

A. Knowledge

Children's understanding of climate change science, impacts and adaptation measures.
Children's understanding of climate change science.
Children's understanding of the impacts of climate change on their families, schools, and their communities.
Children's ability to identify adaptation measures that are relevant to their families, schools and communities.

B. Advocacy

Advocacy by children about climate change adaptation.

Children's communication to their families and schools about vulnerabilities, hazards, and adaptation measures relevant to their community, in ways appropriate to their audiences.

Receptiveness of families, schools, and others in the community to children's voices on climate change adaptation.

C. Policy and Practice

Influence of children on climate change adaptation practice and policy.

The influence of children's perspectives on climate change adaptation practices undertaken by themselves, their families, schools, and the community.

Local government and community leaders' provision of opportunities for children to participate in CCAR planning.

Barangay officials' development and prioritization of policies, ordinance and budgets, based on children's perspectives.

Recognizing that power imbalances between children and adults can affect children's participation, the main FGDs included children only. Adults (other than the research team and local facilitators) were generally not present during discussions. Focus groups were conducted with children by age group (10–12-year-olds, and 13–17-year-olds). These were supplemented with interviews and group discussions with parents, teachers, and local government officials.

The FGDs and the field guide for facilitators were codesigned by the project-implementing partners and conducted by local facilitators familiar to the children to allow frank and open discussions with the children and youth. Recognizing the limitations of self-assessment of knowledge and understanding (Falchikov & Boud, 1989; Sitzmann, Ely, Brown, & Bauer, 2010), the FGDs were not based on children's self-ratings, but posed open questions designed to probe understanding of climate change and perspectives on communication, advocacy, and action.

An important element of analyzing the evidence from the FGDs was to include structured debriefing sessions immediately following each FGD involving implementing and research partners to capture reflections and jointly commence the analytical process against themes. The debriefing session was a guided process whereby project partners discussed and made quantitative assessments against established scales against the indicators (e.g., levels and distribution of CCAR knowledge of children).

Findings: The indicators

The FGDs revealed how children have benefited from the project activities against each of the indicators. For example, some children could describe how climate change affects the interrelated natural and human systems locally and across the Philippines, showing an increase in knowledge relative to the project's baseline assessments. As noted by one project implementer, "They [children and youth] are more articulate than the adults at explaining the impacts of climate change."

Children also shared how they successfully communicated with their families—"I discuss climate change with my family and encourage them to participate in CCA activities" and friends—"We talk about how hot it is and this leads to talking about climate change." Adult respondents noted that children's participation in radio broadcasting and theatre and music productions was a powerful medium for communication. Children highlighted some of the challenges about communication: "Some children won't listen to us if they are in older years;" and also identified ways to overcome these challenges, including through indirect communication: "If it's hard to talk to adults [directly], we use other mediums like posters and video documentaries."

The CCCBA facilitates specific pathways for children to participate in local government decision-making, such as through CVCAs, and also supports children to build confidence to advocate in general. Several children noted their reluctance to communicate directly with local leaders— "Because they are adults, they would know more already." However, several have advocated for action on climate change: "As president of the Youth Association, I proposed to the Barangay Captain to have a meeting on climate change," and "At the Symposium [CCCBA activity] I asked the mayor, 'So what's the next move by local government for adaptation?'" Children shared that they consider local governments' role in adaptation is to regulate building standards, land use and waste management. One child from northern Samar noted, in regards to preventing landslides: "Stop illegal logging. . . If I see someone carrying a chainsaw today, I want to see them carrying seedlings tomorrow." The evidence aligned with Plan's approach to advocacy, illustrating that working with duty bearers (i.e., the government agencies and adult community members who have legal or moral obligations to respect, protect, and fulfill the rights of children) is crucial to create pathways for children's voices to be heard.

The indicators highlighted changes that have occurred by and for children in specific locations; and can be used for general comparisons between locations (although they were not intended for aggregation). The change indicators provided a systematic approach for project implementers to translate multiple, qualitative perspectives and experiences from children and youth into a structured framework. The influence of the CCCBA on project participants has occurred through multiple pathways, actors, and scales. The structure of the indicator set enabled the team to navigate the complexity of these changes and to generate, structure, analyze, and communicate evidence about the details of changes that occurred in practice.

Findings: The Process

The iterative nature of developing and refining the indicators ensured that the process was effective in gauging children's knowledge and perspectives about how the CCCBA has supported adaptation. The flexible approach of integrating research within the project's implementation timeline was valued by project implementers for enabling periodic reflection (cf., e.g., Fisher, Dinshaw, McGray, Rai, & Schaar, this issue). Key process findings are summarized in Table 7.1.

The research acknowledges attribution limitations and sought to assess the contribution of project activities to outcomes. Children in focus groups were asked to reflect on the project activities and to identify sources of information and ideas, from within the project or otherwise. Project implementers provided complementary information about activity details, assessed whether children were accurately recounting what they could have

Table 7.1. Summary of Key Process Findings

Issue	Findings
Assessing contribution	Ways to address contribution include directly asking FGD participants about sources of knowledge, triangulation through interviewing other stakeholders, and project staff providing relevant information.
Impartiality versus familiarity	Children's responses may be influenced by their familiarity with facilitators. On balance, this familiarity was considered essential, because potentially distressing climate disasters would be discussed and facilitators could sensitively navigate discussions with children. Facilitators' awareness of dominant versus quiet children also helped manage FGD interactions. Facilitators and observers noted where and how they considered familiarity to have biased the discussions.
Analysis of qualitative evidence	Commencing analysis immediately following FGDs through a structured debrief session involving facilitators and observers proved effective and efficient. More detailed analysis is best undertaken by someone present at the FGD because of their understanding of the context. However, a third party can undertake analysis with the use of the field notes and process guide.
Replicability	The direct involvement of project implementers in partnership with researchers to develop the process and detailed guidance documents was essential to support practical replicability of the process in other projects. This was facilitated by integrating the research throughout the project implementation timeline.

learned from the project, and subsequently informed the revision of the focus-group questions to elicit increased contribution. Supplementary interviews of stakeholders helped to identify the relative exposure of beneficiary groups to other sources of information about climate change adaptation and advocacy. For example, in one location, teachers confirmed that prior to the program, climate change was only featured in the curriculum within the science subject, which did not address social implications or local adaptation responses.

Involving facilitators who are familiar with the stakeholders can raise concerns in terms of whether the findings will be impartial (House, 2005). In this research, children were generally very familiar with the FGD facilitators, who were CCCBA program officers. Despite some initial concern about how this would influence children's responses, on balance, the facilitators' familiarity with the children, the project, and the research and evaluation processes were considered key and essential to employing culturally appropriate methods and language (including local dialects) to facilitate dialogue with children successfully. Our experience shows that children would have been uncomfortable with unfamiliar adults, and indeed a few of the younger children were initially reluctant to speak in the early parts of the FGD because of the presence of adults other than the facilitators. The pre-existing relationships between facilitators and children were also essential to ensure that discussions about children's experiences of and fears about climate disasters did not cause them to become distressed. Facilitators and observers noted where they considered that children's responses had been affected by their familiarity with the facilitator, or indeed biased by other participants. These observations were noted in the debrief session and taken into account when evaluating the evidence against the indicators.

The debrief sessions are integral to commencing the analysis of evidence against indicators, and the subsequent analysis is ideally undertaken by one of the facilitators or observers present at the relevant FGD. Although evidence is generated from qualitative modes of inquiry, the analysis also includes scalar translations against indicators to help meet quantitative reporting requirements. The scale development and analysis process is currently being finalized, with an eye to address issues of reliability and validity (DeVellis, 2012), as well as to ensure consistency of interpretation and rating. It is also intended that the FGD and analytical process will be used for other projects, and scales are being developed so that FGDs conducted at the commencement of a project can help inform baseline establishment.

The knowledge, advocacy, and practice and policy ToC underpins other child-centered and community development projects implemented by Plan and Save the Children and implementers intend to use this process in the evaluation of child-centered projects in other countries in different contexts. Additionally, the process could also be translated to community development projects (e.g., water and sanitation, livelihoods, health, education) seeking to evaluate change under uncertainty.

Conclusion

Building resilience and adapting to climate change requires many inter-linked changes. The process of developing and gathering evidence from children's and community voices against indicators has revealed many lessons relevant to the project and beyond. In particular, it has illuminated the specific need to work with duty bearers to create pathways for children's voices to be heard, and to navigate power structures that can impede children's advocacy.

Our process was deliberately designed to be integrated within the course of project implementation. This integration has been particularly valuable in terms of tailoring the methods to children and providing a streamlined way for project implementers to gather lessons. The FGDs, stakeholder interviews, analytical approaches, and indicators were designed so that subsequent data collection and analysis could be easily replicated by project implementers themselves; the processes and approaches to participatory FGDs and interviews to inform indicators could also be utilized in independent evaluations. The practice-based design process also demonstrates how applied research organizations can work directly in collaboration with implementing agencies to support learning (Eikeland, 2012; Treichel, Chong, & Gero, 2015).

Children and youth are integral to community-based adaptation efforts, and monitoring and evaluation of programs centered on children, youth, or other vulnerable groups is vital to tracking and understanding how to best support adaptation. By pairing qualitative modes of inquiry that were familiar to implementers and appropriate to the context and participants, with the structure of an indicator set, our research has navigated the complexity of understanding how a community-based program centered on children and youth has created change toward CCAR.

References

Arnold, M., Mearns, R., Oshima, K., & Prasad, V. (2014). *Climate and disaster resilience: The role for community-driven development.* Washington, DC: World Bank. Retrieved from http://documents.worldbank.org/curated/en/2014/02/19127194/climate-disaster-resilience-role-community-driven-development-cdd

Back, E., Cameron, C., & Tanner, T. (2009). *Children and disaster risk reduction: Taking stock and moving forward.* Brighton, United Kingdom: Institute of Development Studies (IDS). Retrieved from http://www.savethechildren.org.uk/sites/default/files/docs/Child_Led_DRR_Taking_Stock_1.pdf

Bours, D., McGinn, C., & Pringle, P. (2014a). *Guidance note 1: Twelve reasons why climate change adaptation M&E is challenging.* Phnom Penh, Cambodia: SEA Change CoP, and Oxford, United Kingdom: UKCIP. Retrieved from http://www.ukcip.org.uk/wordpress/wp-content/PDFs/MandE-Guidance-Note1.pdf

Bours, D., McGinn, C., & Pringle, P. (2014b). *Monitoring & evaluation for climate change adaptation and resilience: A synthesis of tools, frameworks and approaches* (2nd ed.). Phnom Penh, Cambodia: SEA Change CoP, and Oxford, United

Kingdom: UKCIP. Retrieved from http://www.ukcip.org.uk/wordpress/wp-content/PDFs/SEA-Change-UKCIP-MandE-review-2nd-edition.pdf

Brooks, N., Anderson, S., Ayers, J., Burton, I., & Tellam, I. (2011). *Tracking adaptation and measuring development* (Climate Change Working Paper No. 1). London, United Kingdom: International Institute for Environment and Development (IIED). Retrieved from http://pubs.iied.org/pdfs/10031IIED.pdf

CARE. (2010). *Framework of milestones and indicators for community-based adaptation.* London, United Kingdom: Author. Retrieved from http://www.careclimatechange.org/files/toolkit/CBA_Framework.pdf

Child Protection Working Group (CPWG). (2012). *Minimum standards for child protection in humanitarian action.* Geneva, Switzerland: Author. Retrieved from http://cpwg.net/minimum-standards/

Department for International Development (DFID). (2010). *Youth participation in development: A guide for development agencies and policy makers.* London, United Kingdom:DFID Civil Society Organizations (DFID-CSO) Youth Working Group. Retrieved from http://www.youtheconomicopportunities.org/sites/default/files/uploads/resource/6962_Youth_Participation_in_Development.pdf

DeVellis, R. (2012). *Scale development: Theory and applications.* Thousand Oaks, CA: Sage Publications.

Fisher, S., Dinshaw, A., McGray, H., Rai, N., & Schaar, J. (2015). Evaluating climate change adaptation: Learning from methods in international development. *New Directions for Evaluation, 147*, this issue.

Eikeland, O. (2012). Action research: Applied research, intervention research, collaborative research, practitioner research, or praxis research? *International Journal of Actions Research, 8*(1), 9–44.

Falchikov, N., & Boud, D. (1989). Student self-assessment in higher education: A meta-analysis. *Review of Educational Research, 59*(4), 395–430.

Hawrylyshyn, K., Elegado, E., & Tribunalo, B. (2011). *A child centered approach to climate smart disaster risk management.* Working, United Kingdom: Plan International, and Brighton, United Kingdom: Strengthening Climate Resilience (SCR).

House, E. R. (2005). Deliberative democratic evaluation. In S. Mathison (Ed.), *Encyclopedia of evaluation* (pp. 105–109). Thousand Oaks, CA: Sage Publications.

Intergovernmental Panel on Climate Change (IPCC). (2014). *Climate change 2014: Impacts, adaptation and vulnerability* (WGII AR5 Technical summary. Intergovernmental Panel on Climate Change [IPCC]). Cambridge, United Kingdom: Cambridge University Press. Retrieved from http://www.ipcc.ch/report/ar5/wg2/docs/WGIIAR5-IntegrationBrochure_FINAL.pdf

Lamhauge, N., Lanzi, E., & Agrawala, S. (2012). *Monitoring and evaluation for adaptation: Lessons from development co-operation agencies.* (OECD Environment Working Paper No. 38). Paris, France: Organisation for Economic Co-operation and Development (OECD) Publishing. Retrieved from http://www.oecd-ilibrary.org/environment/monitoring-and-evaluation-for-adaptation-lessons-from-development-co-operation-agencies_5kg20mj6c2bw-en

Landsdown, G. (2006). International developments in children's participation: Lessons and challenges. In K. Tisdall, J. Davis, A. Prout, & M. Hill (Eds.), *Children, young people and social inclusion: Participation for what?* Bristol, United Kingdom: Policy Press.

Lloyd, S. J., Kovats, R. S., & Chalabi, Z. (2011). Climate change, crop yields, and under-nutrition: Development of a model to quantify the impact of climate scenarios on child under-nutrition. *Environmental Health Perspectives, 119*(12), 1817–1823. Retrieved from http://www.ncbi.nlm.nih.gov/pubmed/21844000

Massey, I. T. (2011). A proposed model for the analysis and interpretation of focus groups in evaluation research. *Evaluation and Program Planning, 34*, 21–28.

Measham, T. G., Preston, B. L., Smith, T. F., Brooke, C., Gorddard, R., Withycombe, G., & Morrison, C. (2011). Adapting to climate change through local municipal planning: Barriers and challenges. *Mitigation and Adaptation Strategies for Global Change*, 16(8), 889–909. Retrieved from http://link.springer.com/article/10.1007%2Fs11027–011–9301–2

Mitchell, P., & Borchard, C. (2014). Mainstreaming children's vulnerabilities and capacities into community-based adaptation to enhance impact. *Climate and Development*, 6(4), 372–381.

Patton, M. Q. (2015). *Qualitative research and evaluation methods: Integrating theory and practice*. Thousand Oaks, CA: Sage Publications.

Perera, F. P. (2008). Children are likely to suffer most from our fossil fuel addiction. *Environmental Health Perspectives*, 116, 987–990. Retrieved from http://www.ncbi.nlm.nih.gov/pmc/articles/PMC2516589/

Philippine Atmospheric, Geophysical and Astronomical Services Administration (PAGASA). (2011). Climate change in the Philippines. Manila, Philippines: Author. Retrieved from http://pagasa.dost.gov.ph/climate-agromet/climate-change-in-the-philippines

Risdell, J., & McCormick, C. (2013). *Protect my future: The links between child protection and disasters, conflict and fragility*. Working, United Kingdom: Plan International, and London, United Kingdom: Save the Children. Retrieved from http://resourcecentre.savethechildren.se/sites/default/files/documents/7258.pdf

Sheffield, P. E., & Landrigan, P. J. (2011). Global climate change and children's health: Threats and strategies for prevention. *Environmental Health Perspectives*, 119(3), 291–298.

Shier, H. (2001). Pathways to participation: Openings, opportunities and obligations. *Children & Society*, 15(2), 107–117.

Sitzmann, T., Ely, K., Brown, K. G., & Bauer, K. R. (2010). Self-assessment of knowledge: A cognitive learning or affective measure? *Academy of Management Learning & Education*, 9(2), 169–191.

Stewart, D. W., & Shamdasani, P. N. (2015). *Focus groups: Theory and practice* (3rd ed.). Thousand Oaks, CA: Sage Publications.

Tanner, T. (2010). Shifting the narrative: Child-led responses to climate change and disasters in El Salvador and the Philippines. *Children & Society*, 24(4), 339–351.

Tanner, T., Lazcano, J., Lussier, K., Polack, E., Oswald, K., Sengupta, A., ... Rajabali, F., (2010). *Children, climate change and disasters: An annotated bibliography*. Children in a Changing Climate Research. Brighton, United Kingdom: Institute of Development Studies (IDS). Retrieved from http://opendocs.ids.ac.uk/opendocs/handle/123456789/2373#.VOZRHCwwosQ

Treichel, P., Chong, J., & Gero, P. (2015). *A partnership for learning, reflection and evaluation in action: Exploring opportunities for understanding program impact*. Sydney, Australia: Australian Council for International Development (ACFID) University Network Case Study Series. Australian Council for International Development (ACFID). Retrieved from http://www.acfid.asn.au/get-involved/acfid-university-network/files/isf-and-plan-fnl

United Nations Development Programme (UNDP). (2014). *Human development report 2014—Sustaining human progress: Reducing vulnerability and building resilience*. New York, NY: Author. Retrieved from http://hdr.undp.org/en/2014-report/download

UNICEF Office of Research. (2014). *The challenges of climate change: Children on the front line*. Florence, Italy: Author. Retrieved from http://www.unicef-irc.org/publications/pdf/ccc_final_2014.pdf

United Nations University—Institute for Environment and Human Security (UNU-EHS). (2014). *World risk report 2014*. Bonn, Germany: Author. Retrieved from http://www.ehs.unu.edu/article/read/world-risk-report-2014

JOANNE CHONG *is a research director at the Institute for Sustainable Futures, University of Technology, Sydney.*

ANNA GERO *is a senior research consultant at the Institute for Sustainable Futures, University of Technology, Sydney.*

PIA TREICHEL *is the program manager Climate Change Adaptation at Plan International's Australia National Office.*

NEW DIRECTIONS FOR EVALUATION • DOI: 10.1002/ev

Leiter, T. (2015). Linking monitoring and evaluation of adaptation to climate change across scales: Avenues and practical approaches. In D. Bours, C. McGinn, & P. Pringle (Eds.), *Monitoring and evaluation of climate change adaptation: A review of the landscape. New Directions for Evaluation, 147,* 117–127.

8

Linking Monitoring and Evaluation of Adaptation to Climate Change Across Scales: Avenues and Practical Approaches

Timo Leiter

Abstract

Monitoring and evaluation (M&E) efforts to prepare for, adjust to, and reduce the impacts of climate change—a process known as adaptation—can help to understand the results of adaptation interventions and better account for progress over time. Information on adaptation is so far typically gathered through either project and program, or national-level climate change adaptation M&E systems with limited connection between them. However, given that adaptation takes place at multiple scales, a complete picture of the adaptation progress can only be established if information from national and subnational levels is combined. The chapter outlines three avenues illustrated by examples from practice on how information on adaptation interventions and evaluation results can be connected across scales in order to improve the evidence base for adaptation planning and decision making. © 2015 Wiley Periodicals, Inc., and the American Evaluation Association.

The views expressed in this article are those of the author and do not necessarily reflect the views of the Deutsche Gesellschaft für Internationale Zusammenarbeit (GIZ) or of the Federal Ministry for Economic Cooperation and Development (BMZ).

A s the scientific evidence of climate change has continued to strengthen, and as its observable impacts have become more apparent around the globe (Intergovernmental Panel on Climate Change [IPCC], 2013, 2014), efforts to prepare for, adjust to, and reduce the adverse consequences of climate change have become a priority for governments, international organizations, the private sector, and civil society alike. The process of "adjustment in natural or human systems in response to actual or expected climate stimuli or their effects, which moderates harm or exploits beneficial opportunities" is known as adaptation (IPCC, 2001, p. 982). Monitoring and evaluating adaptation is important to understanding what works well in which contexts. Evaluation research informs planning and decision-making, in addition to accountability for climate financing (McKenzie Hedger et al., 2008; Preston, Westway, Dessai, & Smith, 2009).

A particular purpose of the monitoring and evaluation of climate change adaptation interventions is to track progress in adaptation over time and to assess whether countries are actually becoming less vulnerable to the adverse effects of climate change (Ford, Berrang-Ford, Lesnikowski, Barrera, & Heymann, 2013). Such an assessment needs to account for adaptation interventions at local as well as national levels by a variety of actors. However, project- or program-based adaptation M&E systems typically report to the particular funding source only. On the other hand, emerging national-level adaptation M&E systems often merely focus on the implementation of national plans or strategies without taking subnational adaptation actions into account (Hammil, Dekens, Olivier, Leiter, & Klockemann, 2014). This article explores how to connect information on adaptation from subnational- and national-level adaptation M&E systems to get a more complete picture of the adaptation progress. It is based on a review of available literature as well as the experience of the author in working with government agencies on designing adaptation M&E systems. The chapter begins with a brief outline of the challenges of evaluating adaptation, followed by a review of the consideration of scales in countries' current adaptation M&E systems. It then presents three avenues as to how adaptation M&E systems can be connected across scales and discusses how evaluators can implement them in practice.

Assessing Adaptation Across Scales

Adaptation to climate change can occur in a variety of ways, including through policies and planning, capacity building, physical or social measures, and behavior change at different spatial and temporal scales and in virtually every sector (IPCC, 2014, p. 844–849). As a result, adaptation does not have a common reference metric to measure success (IPCC, 2014, p. 853). Instead, successful adaptation needs to be defined in a particular context. Traditional development indicators like "amount of agricultural yields" or "level of income" may indicate successful adaptation, but quantifying the exact contribution of adaptation is more difficult. Assessing the effectiveness of adaptation is furthermore complicated by the long time scales over which climate change unfolds and by parallel changes in climatic and socioeconomic baseline conditions (Bours, McGinn, & Pringle, 2014). It has therefore been suggested to use a combination of process-based indicators together with outcome-based indicators (Brooks, Anderson, Ayers, Burton, & Tellam, 2011; Harley, Horrocks, Hodgson, & Van Minnen, 2008).

The concept of scale describes that events taking place at different spatial levels and different moments in time can influence one another (Gibson, Ostrom, & Ahn, 2000). These interdependencies make issues of scale highly relevant for evaluators who seek to understand change, attribute its causes, and produce relevant knowledge products (Bruyninckx, 2009; Kennedy, Balasubramanian, & Crosse, 2009). The term *level* can be used to describe subunits of a particular scale, for example, local, national, and global levels on a governance scale (Gibson et al., 2000). However, the terms *scales* and *levels* are also used interchangeably (cf., e.g., Adger, Arnell, & Tompkins, 2005; Urwin & Jordan, 2008).

Considerations of spatial and temporal scale are "paramount in determining the success of adaptation" (Palutikof et al., 2013, p. 20; cf. e.g., Adger et al., 2005; Wilbanks & Kates, 1999). To get an overall picture of the adaptation progress, information on adaptation actions at subnational and national levels need to be combined. The national level is important because national governments play an essential role in facilitating adaptation through agenda setting, planning, and budgeting (Berkhout, 2005; Cimato & Mullan, 2010). Lamhauge, Lanzi, and Agrawala (2013, p. 241) state that "future research should therefore focus on M&E approaches that enable evaluators to link individual project or program assessments with evaluations at the national level of countries' success in becoming less vulnerable to climate change." A workshop on adaptation M&E organized by the Adaptation Committee under the United Nations Framework Convention on Climate Change (UNFCCC) also focused on this topic, but did not manage to elaborate any approaches on how such a link could be achieved (Adaptation Committee, 2014, p. 4).

Considerations of Scale in Countries' Adaptation M&E Systems

More than a dozen countries are currently drafting national-level adaptation M&E systems, including France, Germany, Kenya, the Philippines, and the United Kingdom (Hammil et al., 2014; Leiter, 2013; see also European Environment Agency, 2014). Their general purpose is to gather information on the implementation of adaptation strategies and actions and on their effectiveness. The specific design, scope, and methods employed differ according to the national policy context (Hammil et al., 2014). For example, the French monitoring system uses output indicators to track the implementation of 240 measures across 20 sectoral or thematic areas, whereas the results-based M&E system of the Philippines' Climate Change Commission employs results chains that specify activities, outputs, and outcomes (Hammil et al., 2014; Reysset, 2014).

Examining the adaptation M&E systems reviewed by Hammil et al. (2014) indicates that issues of scale and cross-scale dynamics are seldom considered. The French and German adaptation M&E systems focus entirely on the national level, and in France, subnational adaptation actions are foreseen to be analyzed in a separate and independent process (Reysset, 2014, p. 289). The adaptation M&E systems of the Philippines and the United Kingdom partly draw on data from subnational levels, and the United Kingdom intends to report spatially disaggregated figures whenever possible. Yet their main focus is on a national program or plan (Hammil et al., 2014). Norway's and Kenya's adaptation M&E systems do address issues of scale to some extent. Instead of an indicator-based M&E system, Norway has institutionalized a knowledge-exchange process through stakeholder forums and regular surveys of municipalities. The results inform the national vulnerability and adaptation assessments (Hammil et al., 2014; for details of the survey see Amundsen, Berglund, & Westskog, 2010). Kenya's adaptation M&E system covers the national and the next closest government level and proposes a set of similar process and outcome-based indicators for both levels (Ministry of Environment and Mineral Resources, 2012).

Avenues and Practical Approaches

In response to the gap in research (Lamhauge, Lanzi, & Agrawala, 2012, 2013) and in practice as outlined above, this article presents three avenues of linking information on adaptation across scales. Each avenue can be operationalized through different approaches that are illustrated with examples from practice. Linking is understood as any type of connection between M&E systems across scales. The term *M&E system* is understood in a broader sense to cover all mechanisms and processes to gather or provide information on the implementation of climate change adaptation interventions and/or assess their results. This broader definition is necessary because specific M&E mechanisms for adaptation are only just beginning to emerge, and few are currently prepared to measure effectiveness in terms of actual reduction of vulnerability (e.g., see the summary of measuring adaptation in IPCC, 2014, pp. 853–857). The three avenues and their practical approaches are presented below, together with a discussion of their respective advantages and disadvantages. An overview is provided in Table 8.1.

Avenue 1: Standardized Metrics Across Scales

Avenue 1 consists of standardized metrics that are applied in a consistent way across scales. A typical approach under this avenue is to define

Table 8.1. Avenues and Practical Approaches to Link Adaptation M&E Systems Across Scales

Avenue	Practical approach	Example
Avenue 1: Linking based on standardized metrics across scales (aggregation).	Standardized set of indicators for application at multiple levels	Climate funds that aggregate results across projects to the national or global level (e.g., Adaptation Fund, 2014)
	Flexible set of standardized indicators that can be selected according to their applicability to the respective context	Adaptation Monitoring and Assessment Tool of the Global Environment Facility (GEF) (GEF, 2012)
Avenue 2: Linking based on level-specific metrics.	Common themes at national level under which information from independent subnational M&E systems is reported	Suggestion for Mexico's national adaptation M&E system (Deutsche Gesellschaft für Internationale Zusammenarbeit GmbH [GIZ], 2015; Ramos et al., 2014)
	National database of adaptation projects supplementing a national-level indicator system	South Africa's database of adaptation projects (Department of Environmental Affairs, 2014)
Avenue 3: Informal links and synthesis across scales.	Synthesizing available evaluation results and reporting them alongside national indicators	Review of Germany's national adaptation strategy (Hammil et al., 2014)
	Informal dialogue across scales informs a national adaptation assessment in a qualitative way	Norway's "learning by doing" assessment of adaptation actions (Hammil et al., 2014)

a set of standard indicators that are measured at a subnational level and then aggregated to higher levels. Standard indicators are being used by international climate funds, such as the Adaptation Fund, to measure the collective performance of their portfolio (Adaptation Fund, 2014). A prerequisite of this approach is that indicators need to have meaning at different levels, which limits their ability to take specific context into account. Chen and Uitto (2014), who analyzed the challenges of aggregating local actions to global results, point out that mechanical aggregation fails to capture important results, in particular on cross-scale dynamics, because the collective outcome of local actions can be more than the sum of their parts. Furthermore, defining indicators that are universally applicable to a broad range of sectors and interventions, as is the case with adaptation, often leads to rather simple common-denominator indicators such as "number of beneficiaries" or "number of tools developed" (cf. adaptation action indicators in GIZ, 2014). To overcome the latter, another approach is to develop a larger set of more specific indicators from which subnational adaptation interventions can select, based on applicability to the respective context. Although providing more flexibility, Chen and Uitto (2014, p. 120) note that some of these indicators may be rarely used and that aggregating only a selection of indicators would therefore underrepresent outcomes at the portfolio level.

Avenue 2: Level-Specific Metrics

In contrast to Avenue 1, Avenue 2 uses context-specific metrics that are not standardized, while still being aligned to a national system. One approach is to define common themes at national level to which subnational actors can report based on independently operated M&E systems. Mexico is currently considering such an approach in order to integrate information from its federal states into a national adaptation M&E system (GIZ, 2015). Seven overarching monitoring themes have been suggested, including ecosystem services, social systems, and production systems, as well as government and social capacities, each of which is specified by a number of subthemes such as health, food security, and access to water under the social-system theme (Ramos, Altamirano, Klockemann, & Alarcón, 2014). The advantage of this approach is to have agreed-upon focus areas for M&E, while providing flexibility on the exact methods and metrics to be used. On the flip side, in case of high variation this approach limits comparability and the ability to quantify progress across scales.

Another approach, employing level-specific metrics, is being tested in South Africa. As part of the national-level adaptation M&E system, the Department of Environmental Affairs (2014) has set up a public database where adaptation projects can be registered. The database could serve as a useful addition to national-level indicators, because it helps tracking which adaptation actions are taking place at different levels and what their purpose is (e.g., Preston et al., 2009, p. 9). However, in its current form the database does not assist in monitoring implementation or evaluating results of projects, as there is neither an incentive to update information over time nor is there a data quality control mechanism in place.

Avenue 3: Informal Links and Synthesis Across Scales

Avenue 3 encompasses loose or less-structured linkages across adaptation M&E systems. One approach is to synthesize relevant information on the implementation of adaptation at multiple scales as part of the reporting of a national adaptation M&E system. For instance, the review of the German adaptation strategy, which is due at the end of 2015, will consist of three parts: (a) a description of the trends of climate impact and adaptation indicators, (b) results from Germany's vulnerability assessment, and

(c) a review report summarizing experiences of the implementation of the adaptation strategy and the effects it has had (Hammil et al., 2014). The latter may be based on sources of information outside of the formal adaptation monitoring system, which offers the advantage of a better-informed assessment by taking findings from different levels and independent evaluations into account. Another approach adopted by Norway is a mix of formal (stakeholder forum and survey results) and informal (exchange among government bodies) elements, which inform national adaptation assessments in a qualitative way (Hammil et al., 2014).

The example of Norway shows that the three avenues are not mutually exclusive. On the contrary, they can be combined to offset each other's limitations. For instance, Germany uses standardized indicators (Avenue 1) in combination with a review of other information sources (Avenue 3). South Africa aims to introduce national indicators (Avenue 1) to accompany its project database (Avenue 2). And Norway produces a qualitative national assessment based on a survey of municipalities (Avenue 2) and lessons captured through a networking process (Avenue 3).

Discussion: Combining Information on Climate Adaptation Across Scales

Tracking adaptation to climate change is important to understand whether progress is being made in preparing for, adjusting to, and reducing the impacts of climate change, as well as to identify how the adaptation process can be further supported (Ford et al., 2013). M&E systems of a particular project or a national adaptation plan may serve different objectives, but combining their insights has the potential to improve adaptation planning and implementation. Pursuing the strategy of more strategically combining adaptation M&E systems is timely because information on adaptation is so far scarce and scattered (Ford et al., 2013), which hampers the opportunity to learn from experience (Preston et al., 2009).

Standardized metrics, as employed in Avenue 1, can be useful to get an overview of basic results and trends (e.g., Kennedy et al., 2009, pp. 44–47), but they have serious limitations, in particular for understanding how adaptation works and what its context specific outcomes are. Hence, participants of the M&E workshop of the Adaptation Committee "agreed that adding up indicators from local level to get an aggregate number is neither necessarily possible nor desirable" (Adaptation Committee, 2014, p. 4). A top-down approach where local indicators are prescribed by national authorities may also be incompatible with competencies in a federal system of government. Evaluators should therefore seek alternative or complementary ways such as those outlined in Avenues 2 and 3 to cater for the context-specific nature of adaptation. Avenue 3 also illustrates that a link between national and sub-national information on adaptation does not need to be based on a formal or quantitative approach. For the purpose of learning, linking could take the form of documenting and sharing lessons in a qualitative way. Evaluators of adaptation projects or programs, and evaluation offices of funding organizations, should therefore make their findings publicly available and contribute to national climate change databases, such as the one in South Africa.

To date, issues of scale and cross-scaled dynamics seem to be underrepresented in adaptation M&E systems (Hammil et al., 2014). Combining data across scales, as intended in Kenya, could help to infer whether actions taken at national level have resulted in actions at lower levels and vice versa and thus contribute to a better understanding of adaptation processes, their barriers, and enablers. Evaluators should therefore advocate for the consideration of scale in the M&E design (Bruyninckx, 2009). Finally, because people in charge of developing adaptation M&E systems may not have an M&E or adaptation background, evaluators advising them should point out that M&E of adaptation needs to begin with an analysis of the adaptation context and with clarifying the purpose of M&E instead of rushing straight to indicator formulation (Leiter, 2013; Olivier, Leiter, & Linke, 2013).

Conclusions

Relying only on national-level indicators to assess the adaptation progress of a country is missing important aspects, because adaptation is a context-specific process taking place across scales. This article has therefore categorized and illustrated multiple approaches on how national adaptation M&E systems can be linked with subnational ones. Such linkages of information, gathered at different levels, are not limited to using standardized indicators, but can also be based on the sharing of qualitative insights. Evaluators and decision makers should consider the importance of scale and cross-scale dynamics in the design of M&E systems to enhance the understanding of adaptation processes. They are also encouraged to support the sharing of information in order to enhance the basis for tracking adaptation to climate change over time.

References

Adaptation Committee. (2014). *Report on the workshop on the monitoring and evaluation of adaptation—from the fifth meeting of the Adaptation Committee*. Bonn, Germany: UNFCCC Secretariat. Retrieved from http://unfccc.int/files/adaptation/cancun_ adaptation_framework/adaptation_committee/application/pdf/ac_me_ws_report_ final.pdf

Adaptation Fund. (2014). *Methodologies for reporting adaptation fund core impact indicators*. Washington, DC: Author. Retrieved from https://www.adaptation-fund.org/sites/ default/files/AF%20Core%20Indicator%20Methodologies.pdf

Adger, W. N., Arnell, N. W., & Tompkins, E. L. (2005). Successful adaptation to climate change across scales. *Global Environmental Change*, 15, 77–86.

Amundsen, H., Berglund, F., & Westskog, H. (2010). Overcoming barriers to climate change adaptation—A question of multilevel governance? *Environment and Planning C: Government and Policy, 28,* 276–289.

Berkhout, F. (2005). Rationales for adaptation in EU climate change policies. *Climate Policy, 5*(3), 377–391.

Bours, D., McGinn, C., & Pringle, P. (2014). *Guidance note 1: Twelve reasons why climate change adaptation M&E is challenging.* Phnom Penh, Cambodia: SEA Change CoP, and Oxford, United Kingdom: UKCIP. Retrieved from http://www.ukcip.org.uk/wordpress/wp-content/PDFs/MandE-Guidance-Note1.pdf

Brooks, N., Anderson, S., Ayers, J., Burton, I., & Tellam, I. (2011). *Tracking adaptation and measuring development (TAMD)* (Climate Change Working Paper No. 1). London, United Kingdom: International Institute for Environmental Development (IIED). Retrieved from http://pubs.iied.org/pdfs/10031IIED.pdf

Bruyninckx, H. (2009). Environmental evaluation practices and the issue of scale. In Birnbaum, M., & Mickwitz, P. (Eds.), *Environmental program and policy evaluation: Addressing methodological challenges. New Directions for Evaluation, 122,* 31–39.

Chen, S., & Uitto, J. I. (2014). Small grants, big impacts. Aggregation challenges. In J. I. Uitto (Ed.), *Evaluating environment in international development* (pp. 105–122). New York, NY: Routledge.

Cimato, F., & Mullan, M. (2010). *Adapting to climate change: Analyzing the role of government.* London, United Kingdom: Department for Environment, Food and Rural Affairs. Retrieved from http://archive.defra.gov.uk/environment/climate/documents/analysing-role-government.pdf

Department of Environmental Affairs. (2014). *South Africa's national climate change response database.* Pretoria, South Africa: Author.

European Environment Agency (2014). *National adaptation policy processes in European countries.* Luxembourg, Luxembourg: Publications Office of the European Union. Retrieved from http://www.eea.europa.eu/publications/national-adaptation-policy-processes

Ford, J. D., Berrang-Ford, L., Lesnikowski, A., Berrera, M., & Heymann, S. J. (2013). How to track adaptation to climate change: A typology of approaches for national-level application. *Ecology and Society, 18*(3), 40.

Gesellschaft für Internationale Zusammenarbeit (GIZ). (2014). *Repository of adaptation indicators: Real case examples from national monitoring and evaluation systems.* Eschborn, Germany Author. Retrieved from https://gc21.giz.de/ibt/var/app/wp342deP/1443/?wpfb_dl=221

Gesellschaft für Internationale Zusammenarbeit (GIZ). (2015). *Mexico: Monitoring and evaluation of the special program on climate change.* Eschborn, Germany: Author. Retrieved from https://gc21.giz.de/ibt/var/app/wp342deP/1443/?wpfb_dl=238

Gibson, C. C., Ostrom, E., & Ahn, T. K. (2000). The concept of scale and the human dimensions of global change: A survey. *Ecological Economics, 32,* 217–239.

Global Environment Facility (GEF). (2012). *Adaptation monitoring and assessment tool.* Washington, DC: Author. Retrieved from http://www.thegef.org/gef/sites/thegef.org/files/documents/document/GEF%20CC%20Adaptation%20Tracking%20Tool%20Guidelines%2027%20June.pdf

Hammil, A., Dekens, J., Olivier, J., Leiter, T., & Klockemann, L. (2014). *Monitoring and evaluating adaptation at aggregated levels: A comparative analysis of ten systems.* Eschborn, Germany: Deutsche Gesellschaft für Internationale Zusammenarbeit (GIZ) GmbH. Retrieved from https://gc21.giz.de/ibt/var/app/wp342deP/1443/wp-content/uploads/filebase/me/me-guides-manuals-reports/GIZ_2014-Comparative_analysis_of_national_adaptation_M&E.pdf

Harley, M., Horrocks, L., Hodgson, N. & Van Minnen, J. (2008). *Climate change vulnerability and adaptation indicators* (ETC/ACC Technical Paper 2008/9). Bilthoven, The

Netherlands: European Topic Centre on Air and Climate Change, European Environmental Agency.

Intergovernmental Panel on Climate Change (IPCC). (2001). *Contribution of working group II to the third assessment report of the Intergovernmental Panel on Climate Change (IPCC)*. Cambridge, United Kingdom: Cambridge University Press.

Intergovernmental Panel on Climate Change (IPCC). (2013). *Climate change 2013: The physical science basis. Contribution of working group I to the fifth assessment report of the Intergovernmental Panel on Climate Change (IPCC)*. Cambridge, United Kingdom: Cambridge University Press.

Intergovernmental Panel on Climate Change (IPCC). (2014). *Climate change 2014: Impacts, adaptation, and vulnerability. Part A: Global and sectoral aspects. Contribution of working group II to the fifth assessment report of the Intergovernmental Panel on Climate Change (IPCC)*. Cambridge, United Kingdom: Cambridge University Press.

Kennedy, E. T., Balasubramanian, H., & Crosse, W. E. M. (2009). Issues of scale and monitoring status and trends in biodiversity. In M. Birnbaum & P. Mickwitz (Eds.), *Environmental program and policy evaluation: Addressing methodological challenges. New Directions for Evaluation, 122*, 41–51.

Lamhauge, N., Lanzi, E., & Agrawala, S. (2012). *Monitoring and evaluation for adaptation: Lessons from development co-operation agencies* (OECD Environment Working Paper No. 38). Paris, France: Organisation for Economic Co-operation and Development (OECD) Publishing. doi:10.1787/19970900

Lamhauge, N., Lanzi, E., & Agrawala, S. (2013). The use of indicators for monitoring and evaluation of adaptation: Lessons from development cooperation agencies. *Climate and Development, 5*(3), 229–241.

Leiter, T. (2013). *Recommendations for adaptation M&E in practice* [Discussion paper]. Eschborn, Germany: Deutsche Gesellschaft für Internationale Zusammenarbeit (GIZ) GmbH. Retrieved from https://gc21.giz.de/ibt/var/app/wp342deP/1443/?wpfb_dl=132

McKenzie Hedger, M., Mitchell, T., Leavy, J., Greeley, M., Downie, A., & Horrocks, L. (2008). *Desk review: Evaluation of adaptation to climate change from a development perspective*. Brighton, United Kingdom: Institute of Development Studies (IDS)/AEA Group. Retrieved from https://www.climate-eval.org/study/evaluation-adaptation-climate-change-development-perspective

Ministry of Environment and Mineral Resources. (2012). *National performance and benefit measurement framework. Section B: Selecting and monitoring adaptation indicators*. Nairobi, Kenya: Ministry of Environment and Mineral Resources, Government of Kenya. Retrieved from http://www.kccap.info/index.php?option=com_phocadownload&view=category&download=312:section-b-selecting-and-monitoring-adaptation-indicators&id=40:national-performance-and-benefit-measurenment

Olivier, J., Leiter, T., & Linke, J. (2013). *Adaptation made to measure: A guidebook to the design and results-based monitoring of climate change adaptation projects* (2nd ed.). Eschborn, Germany: Deutsche Gesellschaft für Internationale Zusammenarbeit (GIZ) GmbH. Retrieved from https://gc21.giz.de/ibt/var/app/wp342deP/1443/wp-content/uploads/filebase/me/me-guides-manuals-reports/GIZ-2013_Adaptation_made_to_measure_second_edition.pdf

Palutikof, J., Parry, M., Stafford Smith, M., Ash, A. J., Boulter, S. L., & Waschka, M. (2013). The past, present and future of adaptation: Setting the context and naming the challenges. In J. Palutikof, S. L. Boulter, A. J. Ash, M. Stafford Smith, M. Parry, M. Waschka, & D. Guitart (Eds.), *Climate adaptation futures*. Hoboken, NJ: John Wiley & Sons.

Preston, B., Westway, R., Dessai, S., & Smith, T. F. (2009, March). *Are we adapting to climate change? Research and methods for evaluating progress*. Paper presented at Greenhouse 2009: Climate change and resources. Perth, Australia: Commonwealth Scientific and Industrial Research Organisation (CSIRO).

Ramos, M. Z., Altamiran, M. A., Klockemann, L., & Alarcón, S. M. (2014). *Identificación de Indicadores para el Monitoreo y Evaluación de la Adaptación al Cambio Climático en México*. Eschborn, Germany: Deutsche Gesellschaft für Internationale Zusammenarbeit (GIZ) GmbH.

Reysset, B. (2014). The French approach to monitoring an adaptation plan. In A. Prutsch, T. Grothmann, S. McCallum, I. Schauser, & R. Swart (Eds.), *Climate change adaptation manual. Lessons learned from European and other industrialized countries* (pp. 288–293). London, United Kingdom: Routledge.

Urwin, K., & Jordan, A. (2008). Does public policy support or undermine climate change adaptation? Exploring policy interplay across different scales of governance. *Global Environmental Change, 18,* 180–191.

Wilbanks, T. J., & Kates, R. W. (1999). Global change in local places: How scale matters. *Climatic Change, 43,* 601–628.

TIMO LEITER is an advisor on climate change adaptation working at GIZ's Competence Centre for Climate Change.

NEW DIRECTIONS FOR EVALUATION • DOI: 10.1002/ev

Roehrer, C., & Kouadio, K. E. (2015). Monitoring, reporting, and evidence-based learning in the Climate Investment Fund's Pilot Program for Climate Resilience. In D. Bours, C. McGinn, & P. Pringle (Eds.), *Monitoring and evaluation of climate change adaptation: A review of the landscape. New Directions for Evaluation, 147,* 129–145.

9

Monitoring, Reporting, and Evidence-Based Learning in the Climate Investment Funds' Pilot Program for Climate Resilience

Christine Roehrer, Kouassi Emmanuel Kouadio

Abstract

Monitoring and evaluation of funded activities is essential to improve the effectiveness of spending and accountability of climate finance, as well as for transparency and learning. This chapter shares early experiences and emerging lessons from developing the Climate Investment Funds' (CIF) Pilot Program for Climate Resilience (PPCR) participatory results-based monitoring and reporting system. It describes the process of developing the PPCR results framework and takes a critical look at its five core indicators. It then discusses the iterative and participatory development process of the PPCR Monitoring and Reporting Toolkit, as well as progress made and challenges encountered in operationalizing the monitoring and reporting system. The purpose of this system is to track progress toward climate-resilient development at the national level and to monitor, report, and learn from the implementation of PPCR activities at country and project level. The chapter also highlights opportunities to complement monitoring and reporting with evidence-based learning. A summary of lessons learned concludes the chapter. © 2015 Wiley Periodicals, Inc., and the American Evaluation Association.

Monitoring and reporting (M&R) on climate resilience initiatives is still relatively new, especially at aggregated levels. Few monitoring and evaluation models or standards exist to guide this challenging work. Characteristics of climate change, including uncertainty, non-linearity of climate change patterns, and long time horizons pose challenges for adaptation program design (Dinshaw, Fisher, McGray, Rai, & Schaar, 2014, p. 10). Many adaptation initiatives are complex, with fundamental uncertainties about the causal relationship between inputs and outcomes. Other challenges include attributing observed change to specific activities, setting baselines and targets, and assessing the effectiveness of long-term adaptation endeavors within short and medium-term evaluation cycles (Bours, McGinn, & Pringle, 2014a, 2014b; Lamhauge, Lanzi, & Agrawala, 2012).

The Climate Investment Funds (CIF) finance US$.8.1 billion in programming as of January 2015 and is currently the largest operational climate change funding window (CIF, 2014d). The CIF allocates financing through four funding windows, including the Pilot Program for Climate Resilience (PPCR), which assists developing countries to integrate climate resilience into development planning. The US$1.2 billion fund, which is implemented through five Multilateral Development Banks (MDBs), is currently the world's largest operational source of climate resilience and adaptation financing. The PPCR focuses on a small number of countries and transactions, and is now active in nine pilot countries as well as two regional programs.

The PPCR's monitoring and reporting system and toolkit (CIF 2014d) reflect the desire to maintain an inclusive and programmatic approach in the implementation of the investment plans. It aims to engage PPCR stakeholder groups to assess progress toward core indicators, share lessons learned, and discuss challenges.

Development of the PPCR Results Framework and Its Five Core Indicators

Climate change threatens to undermine progress toward development goals, particularly in the poorest countries (Office of the High Representative for the Least Developed Countries, Landlocked Developing Countries and Small Island Developing States [UN-OHRLLS], 2009; World Bank, 2010). Monitoring and evaluation will play an important role in ensuring that adaptation funding is used as effectively as possible and that lessons from early investments inform the continual improvement of adaptation interventions from design to implementation (Olivier, Leiter, & Linke, 2013; Spearman & McGray, 2011). Aggregate monitoring and evaluation systems for adaptation are being developed, but relatively few are fully established and operational (Deutsche Gesellschaft für Internationale Zusammenarbeit

GmbH [GIZ], 2014). The PPCR's monitoring and reporting system was one of the first aggregated monitoring and reporting systems to be rolled out for adaptation.

The Climate Investment Funds were formed in 2008. A first draft M&E framework (CIF, 2009) contained a total of 30 indicators, many of which were not specific enough, impractical to measure, and/or lacked relevance. A subsequent version of the PPCR results framework (CIF, 2010) reduced the number of indicators to 22 across multiple levels and were intended to be applied during both midterm and final evaluations. The aim was to produce a framework that would be flexible, yet robust through an iterative process of field testing and review. Data sources were wide ranging and drawn from project monitoring, surveys, qualitative assessments at the (sub-)national country level, and data gathered to support National Development Plans. In the end, the framework was still found to be complex and cumbersome. Some observed an unfortunate tendency to include all indicators that were of interest to constituents, regardless of the feasibility of doing so (ICF International, 2014). Moreover, the use and application of the PPCR results framework proved to be challenging. The ongoing dialogue process with the MDBs showed that most pilot countries did not have the capacity to establish the required M&E system. In 2011 a process was started to simplify the PPCR logic model and results framework.

The revised and final results framework (CIF, 2012) contains only 11 indicators. Five of these are core indicators, measured and tracked across all the PPCR pilot countries at the programmatic level of the investment plan. These core indicators allow country results to be aggregated and synthesized. Core indicators have proven to be useful for communicating the overall performance of the PPCR as a program, in particular to the donors. The remaining six optional indicators as well as other country and project specific indicators may be used, depending on the countries' specific needs and requirements. Figure 9.1 shows the PPCR revised logic model with indicators, with the five core indicators highlighted in grey. Table 9.1 provides a brief overview on the methodology for monitoring and reporting on the five PPCR core indicators.

Monitoring of the five core indicators is—and needs to be—a country-driven process embedded in a logic model and results framework. The logic model demonstrates the impact chain from the project inputs and activities through to project outputs and outcomes (e.g., Taylor-Powell & Henert, 2008; W.K. Kellogg Foundation, 2004), resulting in potential national or international impacts, including long-term transformational impacts. The results framework links the objectives at each level of the results framework with the indicators. It is designed to operate both within existing national monitoring and evaluation systems and within the MDBs' own managing for development results (MfDR) approach, which aims at continuous improvement of the results-related work of the MDBs.

Figure 9.1. PPCR Revised Logical Model and Results Framework

PPCR Revised logic model and results framework

| Global – CIF Final Outcome (15 – 20 yrs) | Improved climate resilient development consistent with other CIF objectives |

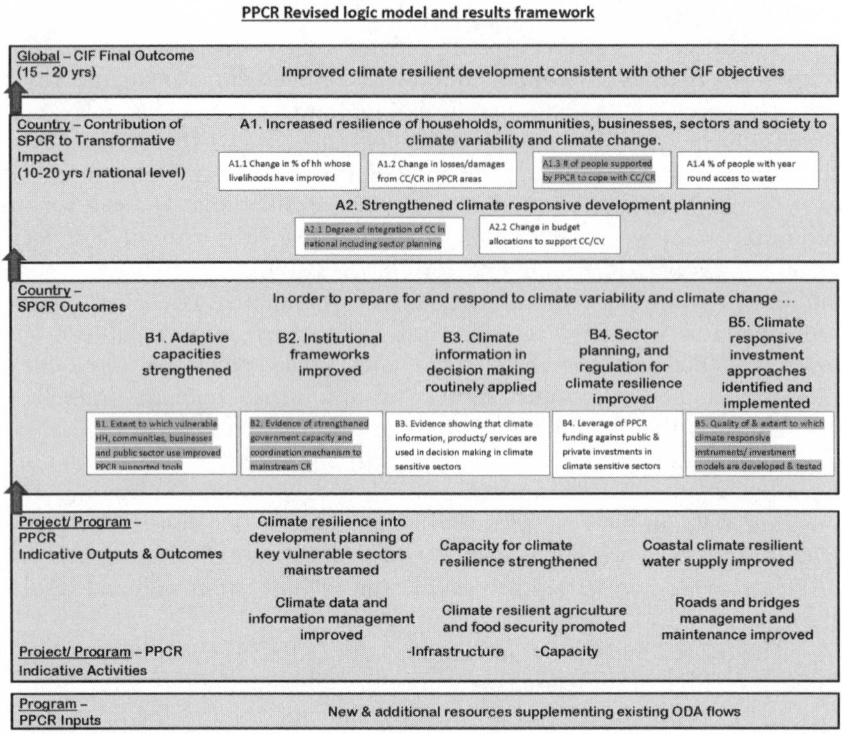

"ANNEX 1: PPCR Revised Logic Model with Indicators" by the Climate Investment Funds (CIF, 2014d, p. 26). Copyright 2014 by the Climate Investment Funds.

Core Indicators 1 and 2 (see Table 9.1) the "degree of integration of climate change in national, including sector, planning" and "evidence of strengthened government capacity and coordination mechanism to mainstream climate resilience," are scored at national level, whereas the data for Core Indicators 3–5 need to be harvested from individual PPCR projects and aggregated at the level of the PPCR investment plan. PPCR monitoring and reporting to the CIF Administrative Unit is at the level of the investment plan only, whereas individual PPCR projects will also report to the implementing MDB.

The PPCR monitoring and reporting toolkit was developed with four key principles, including use of mixed methods, country leadership in M&E, stakeholder engagement, and an iterative process of learning-by-doing. Ultimately, with the use of the final monitoring and reporting toolkit PPCR pilot countries exhibited greater ownership and adherence to the monitoring and reporting processes (CIF, 2014e, pp. 4–5, 30).

Table 9.1. Overview of PPCR Core Indicators, Rationale, and Related Data-Collection Information

#	Core indicator description	Rationale	Type of indicator	Level of data collection	Data-collection instrument
1	Degree of integration of climate change in national, including sector, planning	This indicator is designed to capture the extent to which considerations of climate resilience (risks, opportunities) are integrated into planning processes at national and sectoral levels.	Qualitative/country self-assessment	National level/scoring workshop	Scorecard: scores range from 0 to 10
2	Evidence of strengthened government capacity and coordination mechanism to mainstream climate resilience	This indicator is important to demonstrate that the PPCR's support to pilot country governments results in improved institutions and institutional frameworks for mainstreaming climate resilience.	Qualitative/country self-assessment	National level/scoring workshop	Scorecard: scores range from 0 to 10
3	Quality and extent to which climate responsive instruments/investment models are developed and tested	This indicator estimates (as best as possible) the extent to which the PPCR is identifying and implementing climate responsive investment approaches, by documenting the instruments and models that have been developed and tested with PPCR support and assessing their quality.	Qualitative/project self-assessment	Project level data, aggregate at national level	Scorecard: scores range from 0 to 10

(Continued)

Table 9.1. Continued

#	Core indicator description	Rationale	Type of indicator	Level of data collection	Data-collection instrument
4	Extent to which vulnerable households, communities, businesses, and public sector services use improved PPCR supported tools, instruments, strategies, and activities to respond to climate variability or climate change	This indicator measures the extent to which the PPCR is strengthening the adaptive capacities of target stakeholders in a particular country or region, by measuring their uptake of climate responsive tools, instruments, strategies, and activities that the PPCR is supporting.	Quantitative	Project level data, aggregate at national level	Table: numeric data
5	Number of people supported by the PPCR to cope with the effects of climate change	This indicator determines whether PPCR projects/programs for climate resilience action reach and support people on the ground as intended. Thus it estimates (as best as possible) the number of people directly supported by the PPCR to cope with the effects of climate change in a particular country.	Quantitative	Project level, and aggregate at national level	Table: numeric data

Note. Adapted from "Table 3: Brief overview on the methodology for monitoring and reporting on the five PPCR core indicators" by the Climate Investment Funds (CIF; 2014e, p. 11). Copyright 2014 by the Climate Investment Funds.

Progress and Challenges in Operationalizing the Monitoring and Reporting

Monitoring and reporting on the PPCR with the use of scorecards provides a relatively new approach to the monitoring of climate resilience and adaptation. Stakeholders participate in roundtables to reflect on and rank progress toward indicators. The approach puts as much emphasis on learning as it does on results, both through the scores themselves and through qualitative narratives. Both the journey and the destination are important. This leads not only to more credible data to be reported to the CIF Administrative Unit, but also to stakeholder participation, transparency, accountability, and learning within the PPCR pilot country.

After only 2 years (2013 and 2014) of implementation of the PPCR monitoring and reporting system by the pilot countries in collaboration with the MDBs, initial observations of progress using this approach are highlighted below.

As part of the continued dialogue with PPCR pilot countries and the MDBs, the CIF Administrative Unit learned in 2013 that stakeholder participation that year in scoring indicators 1 and 2 was patchy. The following year, however, stakeholder consultations were organized, and the outcomes were properly documented and shared with the CIF. Some countries conducted their own data analysis and synthesis and submitted them, together with the scorecards and reporting tables.

A successful example is Samoa (CIF, 2014e, p. 21) where a PPCR Climate Resilience Investment Coordination Unit (CRICU) was established in 2011. The experience with CRICU has been so positive that the Ministry of Finance will soon extend the mandate of the unit to coordinate all incoming climate finance for Samoa.

The experience in Cambodia (Royal Government of Cambodia, 2012, p.12) highlighted that it was useful to build on existing national systems and provide additional resources to integrate the PPCR results framework into existing mechanisms.

Although some progress has been made, challenges remain. These include:

1. Limited monitoring and reporting capacity. Building country capacity is critical for the longer-term viability of the PPCR monitoring and reporting system. Although some initiatives have been undertaken by the CIF Administrative Unit to strengthen country capacity and support country reporting, in most cases countries lack internal capacity and rely on external consultants.
2. Stakeholder participation. Although there is a growing number of countries that have actively engaged stakeholders in the second reporting round, some countries have not yet involved all relevant

NEW DIRECTIONS FOR EVALUATION • DOI: 10.1002/ev

stakeholders, in particular civil society organizations and the private sector. Diversity of stakeholders remains a critical challenge.

3. Countries are required to report on multiple climate finance funding streams.

In response to this challenge faced by many countries—for example Nepal's Climate Change Program (CCP), which comprises all climate-change–related projects in the country, including the PPCR, adopted the PPCR five core indicators as the monitoring framework for tracking programmatic and institutional change arising from implementation of the CCP (CIF, 2014e, p. 6).

The following section outlines examples that demonstrate how data collected from PPCR pilot countries has been compiled, aggregated, and analyzed by the CIF.

Core Indicator 1: Degree of Integration of Climate Change Into National and Sector Planning

The spider diagram given in Figure 9.2 shows the progress made by the Commonwealth of Dominica in mainstreaming climate change into its development planning.

Figure 9.2.   Integration of Climate Change into National Planning—Commonwealth of Dominica

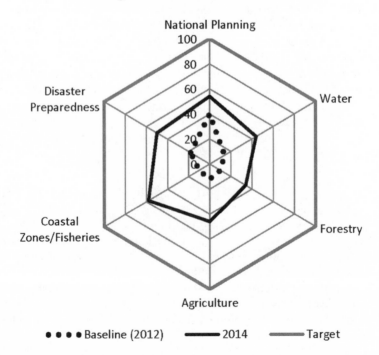

Since the endorsement of the Commonwealth of Dominica's PPCR investment plan in 2012, the level of integration of climate change into development planning has slightly increased from the baseline figure of 40% to 54%. This progress is due to the adoption of strategic policy documents such as: the National Plan for Climate Adaptation Policy, Nationally Appropriate Mitigation Actions, and the Dominica Low Carbon Climate Resilience Strategy 2012–2020 (Government of Dominica, 2013).

On average, as Figure 9.2 demonstrates, integration of climate change in the Commonwealth of Dominica's five priority sectors have moved from mere intention in 2012 to at least recognition. The coastal zone/fisheries sector experienced the greatest increase (29% in 2013 to 54% in 2014). This is partly due to the integration and use of climate screening tools for investments in these sectors (Government of Dominica, 2014).

Core Indicator 2: Evidence of Strengthened Government Capacity and Coordination Mechanism To Mainstream Climate Resilience

The spider diagram in Figure 9.3 is used to show progress made by Saint Lucia in strengthening its government capacity to mainstreaming climate resilience.

At the national government level, government capacity has increased from the baseline (43%) to 48% in 2014 due to improved availability of information to facilitate informed decision-making (CIF 2014e, p. 61–63).

Figure 9.3. Strengthened Government Capacity to Mainstream Climate Resilience—Saint Lucia

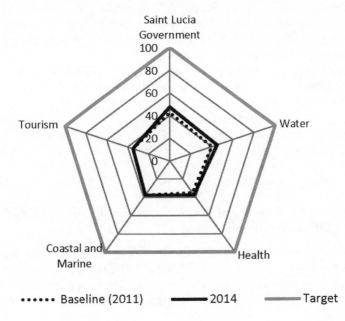

•••••• Baseline (2011)   ▬▬▬ 2014   ▬▬▬ Target

For example, post Hurricane Tomas, a study was conducted assessing all health facilities, schools, and community centers to determine needs in terms of safety and climate resilience. In the water sector, capacity building has increased from 40% to 48% as a result of current mapping of infrastructure, improved capacity of using geographical information systems for building climate resilience, and training on climate change for water resource managers. The health sector also saw improvement (from 35% to 38%) due to training of health personnel on climate change and disaster risk reduction. No changes since the baseline were reported for the coastal and marine or tourism sectors (CIF 2014e)

Core Indicator 3: Quality and Extent to Which Climate Responsive Instruments/Investment Models Are Developed and Tested

Core Indicator 4: Extent to Which Vulnerable Households, Communities, Businesses, and Public-Sector Services Use Improved PPCR-Supported Tools, Instruments, Strategies and Activities To Respond to Climate Variability and Climate Change

Indicators 3 and 4 (above) caused some confusion when the toolkit was piloted in Niger (CIF 2014e, p. 59). People initially struggled to understand the definitions of climate responsive instruments and investments. The improved guidance (CIF, 2014d) now contains more comprehensive and clear definitions.

As shown in Table 9.2, in 2014, 26 approved projects reported on 198 tools/instruments. These tools/instruments fall into five broad categories: analysis and knowledge assets (83), technology and infrastructure (39), public-awareness platforms (38), public and community services (24), and financial instruments (14). Looking at those numbers, it needs to be considered that the 26 currently approved projects only represent a small share of the total PPCR portfolio of 75 projects.

In most of the countries, because of the early-implementation stage of most projects, climate-responsive instruments and investment models have either not been identified or have not been tested yet. Estimates from the so-far-approved 26 projects suggest that over the course of the projects' life cycles, 680,263 households, 4,116 businesses, and 4,699 public-services entities in 5,979 communities are expected to use and benefit from PPCR-supported climate responsive tools/instruments. As more PPCR projects are approved, this number will grow.

Core Indicator 5: Number of People Supported by the PPCR to Cope With the Effects of Climate Change

This indicator determines whether PPCR projects/programs reach and support people on the ground as intended. In 2014, only 10 countries reported on this indicator. This indicator has received some criticism. It was proposed to measure an intervention's success in terms of reduced

Table 9.2. Tools/Investment Models Identified by Countries in Their Results Reports

		PPCR Supported Instruments/Tools per Category					
	Number of approved projects	Financial instruments	Technologies or infrastructure investments	Data, analytical work, technical studies, and knowledge assets	Public/community services	Public awareness platforms	Total number of instruments/tools
Plurinational State of Bolivia	1		0	15	0	1	16
Cambodia	5	3	3	5	2	4	17
Dominica	1		5	6	1	0	12
Grenada	1		4	3	4	3	14
Mozambique	5	2	12	9	7	2	32
Nepal	4	2	8	13	3	4	30
Niger	1		0	3	0	0	3
SVG	1		0	12	1	4	17
Tajikistan	4	5	4	14	4	13	40
Yemen	1		1	0	0	5	6
Zambia	2	2	2	3	2	2	11
Total	26	14	39	83	24	38	198

vulnerability and increased resilience, rather than just the number of people receiving support (Brooks et al., 2013).

Problems remain, including data quality. Some of the pilot countries have not established scoring criteria for the qualitative indicators 1, 2, and 3; hence, scores provided are subjective. Others established their scoring criteria only during the second reporting year but did not revisit and revise their baseline scores. Countries were asked to justify or provide evidence on their scores using examples that support the change in score. All countries did not do this systematically. Finally, the data quality assurance process has not been effective in some pilot countries because of the lack of clarity of the roles and responsibilities of the different actors.

Opportunities To Complement Monitoring and Reporting With Evidence-Based Learning

The term *evidence-based learning* is still relatively new in the context of climate change (Anderson, Khan, Fikreyesus, & Gomes, 2014). For the CIF it spans both formal evaluations and other systematic assessments that use best-available evidence, research, data, and other relevant information (CIF, 2014a) to promote learning that advances climate change mitigation and adaptation investments and results. The concept is informed by the fields of evaluation and evidence-based practice—an interdisciplinary, clinical approach that began with evidence-based medicine (Sackett, Rosenberg, Gray, Haynes, & Richardson, 1996; Sackett, Straus, Richardson, Rosenberg, & Haynes, 2000) and has since been applied in the medical, education, mental health, and lately climate change fields. Evidence-based learning for climate change interventions covers systematic approaches—both quantitative and qualitative, including but not limited to formal evaluations—that inform and improve the design, implementation, and results of climate change interventions (CIF, 2014a).

The CIF is seeking to further improve and complement its M&R system including through:

1. *A stock-taking of current evidence-based learning approaches that are already built into project budget and activities.* Although all MDBs have procedures in place to assure project quality and accountability (cf., e.g., Asian Development Bank [ADB], 2007; African Development Bank [AfDB], 2011; European Bank for Reconstruction and Development [EBRD], 2013; McCarthy, Winters, Linares, & Essam, 2012), what remains unclear is whether these activities can *all* be considered within the rubric of "evidence-based learning."
2. *Development of a menu of suggested approaches to evidence-based learning throughout the project cycle.* This menu (Table 9.3; CIF, 2014a), developed in collaboration with MDB stakeholders, spanned six

Table 9.3. Overview of Suggested Approaches for Evidence-Based Learning Within the PPCR

	Relevant stage of project/program				
Approach	Ex ante	Design[a]	Midcourse	At end	Ex post
1. Adaptive capacity assessment					
2. Developmental evaluation					
3. Formative evaluation					
4. Real-time learning					
5. Rapid stakeholder consultation					
6. Vulnerability assessment					

Note. Adapted from "Table 1: Selected Approach Options for CIF Programs", by the Climate Investment Funds (CIF, 2014a, p. 1). Copyright 2014 by the Climate Investment Funds.
[a]Design refers to the stage after a project has been selected, when details of strategy and implementation plans are being developed.

approaches for the PPCR that range in their objectives, methods and applicability across projects. They apply to different project stages, including before projects begin (ex ante), during implementation (midcourse), and after project completion (ex post). For each approach a fact sheet was developed to help provide a common vocabulary and tool set.

3. *Overview sheet for the PPCR outlining the feasibility of generating evidence of results and related evaluation questions.* A short overview sheet (CIF, 2014c) was developed to guide colleagues' thinking in their portfolio assessment. This sheet contained observations regarding the feasibility of generating evidence for each of the intended results as well as examples of relevant evaluation questions.

4. *Portfolio assessment to identify candidate projects for additional evidence-based learning.* As a first step each MDB undertook a review of their ongoing and planned projects within the CIF portfolio, in order to identify those that would lend themselves to evidence-based learning.

The scoping proposals for the PPCR projects, together with those of the other three programs of the CIF, which outlined the proposed activities and their costs, were taken to the overarching CIF governing body for decision-making in June 2014. At that time it was decided that money allocated for project implementation should not be used for research purposes. The MDBs and bilateral donors were invited to identify resources to co-finance this work. In the subsequent meeting of the CIF governing body one bilateral donor pledged US$9.4 million to support the work on evidence-based learning. Subsequently a feasibility assessment to guide the use of these funds was undertaken in early 2015. Its recommendations are pending, but will help guide the future work of the PPCR in this area.

Summary of Lessons Learned

Key lessons learned throughout the process of designing and implementing the PPCR participatory results-based monitoring and reporting systems follow.

It is evident that an agreed-results framework should always inform the design of investment plans and projects. Given the need to start designing investment plans and projects immediately once the PPCR pledges were received, this was only partly done in the case of the PPCR. Retrofitting core indicators into some existing investment plans and projects required extra efforts. In addition, links between existing indicators at project level with PPCR core indicators at investment-plan level had to be made so that project-level indicator information collected would also inform the core indicators at investment plan level.

Country leadership is essential to ensure effective implementation of the PPCR participatory monitoring system. Building countries' capacities for monitoring and reporting is important in order to make country-level M&E systems sustainable over time. It is important, wherever possible, to reduce complexity of application and implementation of the results framework so that it can be implemented in different contexts with varying levels of country capacity. Straightforward guidance and adequate human and financial resources are a must. In addition, it is important to clarify the roles and responsibilities of all the actors involved in the monitoring and reporting system in order to assure good quality data.

Core indicators are useful for aggregation and synthesis at the level of the investment plan in country and at the level of the global PPCR. Development is context specific, so in addition to the core indicators, countries can develop their own additional indicators of relevance to their situation at project and investment-plan levels. That being said, all core indicators must be included at the level of the investment plan. Countries are invited to use their own methodologies, assumptions, and criteria in implementing the framework, as long as this information is well documented for reasons of transparency, accountability, and building institutional knowledge for program continuity.

Involving the end users, including PPCR pilot-country focal points and their teams, in the development of the toolkit in a participatory and inclusive way helped to ensure the user perspective remained front and center. It enhanced both the quality and the relevance of the toolkit. Those who had been part of the process gained real ownership. Although the core indicators will not change, the toolkit is considered a living document that is continuously evolving, adapting, and improving over time as all stakeholders involved in monitoring and reporting learn from experience.

Routine annual monitoring and reporting is not sufficient to answer key questions as to *why* and *how* certain approaches in implementation of the investment plans have worked better than others. It also does not fully answer key questions as to whether the PPCR is achieving its stated objectives of transformative impact (increased resilience of households, communities, businesses, sectors, and society to climate variability and climate change; strengthened climate responsive development planning) and outcome objectives (adaptive capacities strengthened; adequate institutional frameworks in place; climate information in decision-making routinely applied; improved sector planning and regulation for climate resilience; and climate responsive investment approaches identified and implemented). Because adaptation and resilience building are relatively new fields, learning from experience and feeding those insights back as adjustments to implementation is vitally important. Initiatives that go beyond what is routinely required, such as evidence-based learning, need dedicated additional funding for their implementation.

References

African Development Bank (AfDB). (2011). *Monitoring and evaluation frameworks and the performance and governance of international funds.* Abidjan, Ivory Coast: Author. Retrieved from http://www.afdb.org/fileadmin/uploads/afdb/Documents/Generic-Documents/Monitoring%20and%20evaluation%20frameworks%20and%20the%20performance%20and%20governance%20of%20international%20funds.pdf

Anderson, S., Khan, F., Fikreyesus, D., & Gomes, M. (2014). *Forwards and backwards evidence-based learning on climate adaptation* [Briefing]. London, United Kingdom: International Institute for Environmental Development (IIED). Retrieved from http://pubs.iied.org/pdfs/17257IIED.pdf

Asian Development Bank (ADB). (2007). *Guidelines for preparing a design and monitoring framework.* Manila, Philippines: Author. Retrieved from http://www.adb.org/sites/default/files/institutional-document/32509/files/guidelines-preparing-dmf.pdf

Bours, D., McGinn, C., & Pringle, P. (2014a). *Guidance note 1: Twelve reasons why climate change adaptation M&E is challenging.* Phnom Penh, Cambodia: SEA Change CoP, and Oxford, United Kingdom: UKCIP. Retrieved from http://www.ukcip.org.uk/wordpress/wp-content/PDFs/MandE-Guidance-Note1.pdf

Bours, D., McGinn, C., & Pringle, P. (2014b). *Monitoring & evaluation for climate change adaptation and resilience: A synthesis of tools, frameworks and approaches* (2nd ed.). Phnom Penh, Cambodia: SEA Change CoP, and Oxford, United Kingdom: UKCIP. Retrieved from http://www.ukcip.org.uk/wordpress/wp-content/PDFs/SEA-Change-UKCIP-MandE-review-2nd-edition.pdf

Brooks, N., Anderson, S., Burton, I., Fisher, S., Rai, N., & Tellam, I. (2013). *An operational framework for tracking adaptation and measuring development* (Climate Change Working Paper No. 5). London, United Kingdom: International Institute for Environmental Development (IIED). Retrieved from http://pubs.iied.org/pdfs/10038IIED.pdf

Climate Investment Funds. (2009). *PPCR results framework PPCR/SC.3/6.* Washington, DC: Climate Investment Funds (CIF) administrative unit. Retrieved from http://siteresources.worldbank.org/INTCC/Resources/PPCRResultsFrameworkApril24.pdf

Climate Investment Funds. (2010). *PPCR results framework PPCR/SC.7/4.* Washington, DC: Climate Investment Funds (CIF) administrative unit. Retrieved from https://www.climateinvestmentfunds.org/cif/sites/climateinvestmentfunds.org/files/PPCR%204%20Results%20Framework%20nov2010_1.pdf

Climate Investment Funds. (2012). *Revised Pilot Program for Climate Resilience (PPCR) results framework*. Washington, DC: Climate Investment Funds (CIF) administrative unit. Retrieved from http://www.climateinvestmentfunds.org/cif/sites/climat einvestmentfunds.org/files/Revised_PPCR_Results_Framework.pdf

Climate Investment Funds. (2014a). *Approaches to evidence-based learning throughout the CIF project cycle*. Washington, DC: Climate Investment Funds (CIF) administrative unit. Retrieved from https://www.climateinvestmentfunds.org/cif/sites/climat einvestmentfunds.org/files/CIF_Evidence_Based_Approaches_March2014.pdf

Climate Investment Funds. (2014b). Climate Investment Funds finances [Web page]. Washington, DC: Climate Investment Funds (CIF) administrative unit. Retrieved from http://www.climateinvestmentfunds.org/cif/finances

Climate Investment Funds. (2014c). Pilot Program for Climate Resilience (PPCR): Feasibility of generating evidence of results and related questions. Washington, DC: Climate Investment Funds (CIF) administrative unit. Retrieved from https://climateinvestmentfunds.org/cif/sites/climateinvestmentfunds.org/files/PPCR_ evidence_questions_April2014.pdf

Climate Investment Funds. (2014d). *Pilot Program for Climate Resilience (PPCR) monitoring and reporting toolkit*. Washington, DC: Climate Investment Funds (CIF) administrative unit. Retrieved from https://www.climateinvestmentfunds.org/ cif/node/14652

Climate Investment Funds. (2014e). *2014 Pilot Program for Climate Resilience (PPCR) results report*. Washington, DC: Climate Investment Funds (CIF) administrative unit. Retrieved from https://www.climateinvestmentfunds.org/cif/content/ ppcr-results-report

Deutsche Gesellschaft für Internationale Zusammenarbeit GmbH (GIZ). (2014). *Monitoring and evaluating adaptation at aggregated levels: A comparative analysis of ten systems*. Eschborn, Germany: Author. GmbH. Retrieved from https://gc21.giz.de/ibt/ var/app/wp342deP/1443/wp-content/uploads/filebase/me/me-guides-manuals-reports /GIZ_2014-Comparative_analysis_of_national_adaptation_M&E.pdf

Dinshaw, A., Fisher, S., McGray, H., Rai, N., & Schaar, J. (2014). *Monitoring and evaluation of climate change adaptation: Methodological approaches* (OECD Environment Working Papers No. 74). Paris, France: Organisation for Economic Co-operation and Development (OECD) Publishing. Retrieved from http://dx.doi.org/ 10.1787/5jxrclr0ntjd-en

European Bank for Reconstruction and Development (EBRD). (2013). *Evaluation policy*. London, United Kingdom: Author. Retrieved from http://www.ebrd.com/cs/Satellite? c=Content&cid=1395241631988&pagename=EBRD%2FContent%2FDownloadD ocument

Government of Dominica. (2013). *Dominica low carbon climate-resilient development strategy 2012–2020*. Roseau, Commonwealth of Dominica: Author. Retrieved from http://www4.unfccc.int/sites/nama/_layouts/UN/FCCC/NAMA/Download.aspx?List Name=NAMA&Id=34&FileName=dominica_low_carbon_climate_resilient_strateg y__(finale)%5B1%5D.pdf

Government of Dominica. (2014). *Dominica 2014 PPCR results report*. Washington, DC: Climate Investment Funds (CIF) Administrative Unit. Retrieved from https://www.climateinvestmentfunds.org/cif/node/17124

ICF International (2014). *Independent evaluation of the Climate Investment Funds (CIF)* [Evaluation Report]. Washington, DC: World Bank. Retrieved from http:// ieg.worldbank.org/Data/reports/cif_eval_final.pdf

Lamhauge, N., Lanzi, E., & Agrawala, S. (2012). *Monitoring and evaluation for adaptation: Lessons from development co-operation agencies* (OECD Environment Working Paper No. 38). Paris, France: Organisation for Economic Co-operation and Development (OECD) Publishing. doi: 10.1787/19970900

McCarthy, N., Winters, P., Linares, A. M., & Essam, T. (2012). *Indicators to assess the effectiveness of climate change projects.* Washington, DC: Inter-American Development Bank (IDB). Retrieved from http://publications.iadb.org/bitstream/handle/11319/5447/Indicators%20to%20Asses%20the%20Effectiveness%20of%20Climate%20Change%20Projects.pdf

Office of the High Representative for the Least Developed Countries, Landlocked Developing Countries and Small Island Developing States (UN-OHRLLS) (2009). *The impact of climate change on the development prospects of the least developed countries and small island developing states.* New York, NY: Author. Retrieved from http://unohrlls.org/custom-content/uploads/2013/11/The-Impact-of-Climate-Change-on-The-Development-Prospects-of-the-Least-Developed-Countries-and-Small-Island-Developing-States1.pdf

Olivier, J., Leiter, T., & Linke, J. (2013). *Adaptation made to measure: A guidebook to the design and results-based monitoring of climate change adaptation projects* (2nd ed.) [Manual]. Eschborn, Germany: Deutsche Gesellschaft für Internationale Zusammenarbeit (GIZ) GmbH. Retrieved from https://gc21.giz.de/ibt/var/app/wp34
2deP/1443/wp-content/uploads/filebase/me/me-guides-manuals-reports/GIZ-2013_
Adaptation_made_to_measure_second_edition.pdf

Royal Government of Cambodia. (2012). *Towards an effective monitoring and evaluation framework for adaptation to climate change in Cambodia.* Booklet published as a result of the Strategic Program for Climate Resilience Climate Change Adaptation Monitoring and Evaluation Workshop 4–5 October 2012, Phnom Penh Hotel, Phnom Penh, Cambodia. Phnom Penh, Cambodia: Author, and Manila, Philippines: Asian Development Bank (ADB), and Washington, DC: World Bank.

Sackett, D. L., Rosenberg, W. M., Gray, J. A., Haynes, R. B., & Richardson, W. S. (1996). Evidence based medicine: What it is and what it isn't. *British Medical Journal, 312*(7023), 71–72.

Sackett, D. L., Straus, S. E., Richardson, W. S., Rosenberg, W. M., & Haynes, R. B. (2000). *Evidence-based medicine: How to practice and teach EBM.* London, United Kingdom: Churchill Livingstone.

Spearman, M., & McGray, H. (2011). *Making adaptation count: Concepts and options for monitoring and evaluation of climate change adaptation* [Manual]. Eschborn, Germany: Deutsche Gesellschaft für Internationale Zusammenarbeit (GIZ) GmbH, and Bonn, Germany: Bundesministerium für wirtschaftliche Zusammenarbeit und Entwicklung (BMZ), and Washington, DC: World Resources Institute (WRI). Retrieved from http://www.wri.org/publication/making-adaptation-count

Taylor-Powell, E., & Henert, E. (2008). *Developing a logic model: Teaching and training guide.* Madison, WI: University of Wisconsin, Extension Cooperative Extension Program Development and Evaluation. Retrieved from http://www.uwex.edu/ces/pdande/evaluation/pdf/lmguidecomplete.pdf

W.K. Kellogg Foundation (2004). *Using logic models to bring together planning, evaluation, and action: Logic model development guide.* Battle Creek, MI: Author. Retrieved from http://www.smartgivers.org/uploads/logicmodelguidepdf.pdf

World Bank. (2010). *World development report 2010: Development and climate change.* Washington, DC: Author. Retrieved from http://go.worldbank.org/UVZ0HYFGG0

CHRISTINE ROEHRER *is the Climate Investment Funds' (CIF), USA, senior monitoring and evaluation specialist.*

KOUASSI EMMANUEL KOUADIO *is the Climate Investment Funds' (CIF), USA, monitoring and evaluation specialist.*

NEW DIRECTIONS FOR EVALUATION • DOI: 10.1002/ev

INDEX

147

ORDER FORM SUBSCRIPTION AND SINGLE ISSUES

DISCOUNTED BACK ISSUES:

Use this form to receive 20% off all back issues of *New Directions for Evaluation*.
All single issues priced at **$23.20** (normally $29.00)

TITLE	ISSUE NO.	ISBN

Call 1-800-835-6770 or see mailing instructions below. When calling, mention the promotional code JBNND to receive your discount. For a complete list of issues, please visit www.josseybass.com/go/ev

SUBSCRIPTIONS: (1 YEAR, 4 ISSUES)

☐ New Order ☐ Renewal

U.S.	☐ Individual: $89	☐ Institutional: $358
CANADA/MEXICO	☐ Individual: $89	☐ Institutional: $398
ALL OTHERS	☐ Individual: $113	☐ Institutional: $432

Call 1-800-835-6770 or see mailing and pricing instructions below.
Online subscriptions are available at www.onlinelibrary.wiley.com

ORDER TOTALS:

Issue / Subscription Amount: $ _____

Shipping Amount: $ _____
(for single issues only – subscription prices include shipping)

Total Amount: $ _____

SHIPPING CHARGES:

First Item	$6.00
Each Add'l Item	$2.00

(No sales tax for U.S. subscriptions. Canadian residents, add GST for subscription orders. Individual rate subscriptions must be paid by personal check or credit card. Individual rate subscriptions may not be resold as library copies.)

BILLING & SHIPPING INFORMATION:

☐ **PAYMENT ENCLOSED:** *(U.S. check or money order only. All payments must be in U.S. dollars.)*

☐ **CREDIT CARD:** ☐ VISA ☐ MC ☐ AMEX

Card number _____ Exp. Date_____

Card Holder Name_____ Card Issue # _____

Signature _____ Day Phone_____

☐ **BILL ME:** *(U.S. institutional orders only. Purchase order required.)*

Purchase order # _____
Federal Tax ID 13559302 • GST 89102-8052

Name_____
Address_____
Phone_____ E-mail_____

Copy or detach page and send to: **John Wiley & Sons, One Montgomery Street, Suite 1000, San Francisco, CA 94104-4594**

Order Form can also be faxed to: **888-481-2665**

PROMO JBNND